The Cambridge Manuals of Science and
Literature

THE ATMOSPHERE

The Hon. Henry Cavendish (1731—1810).

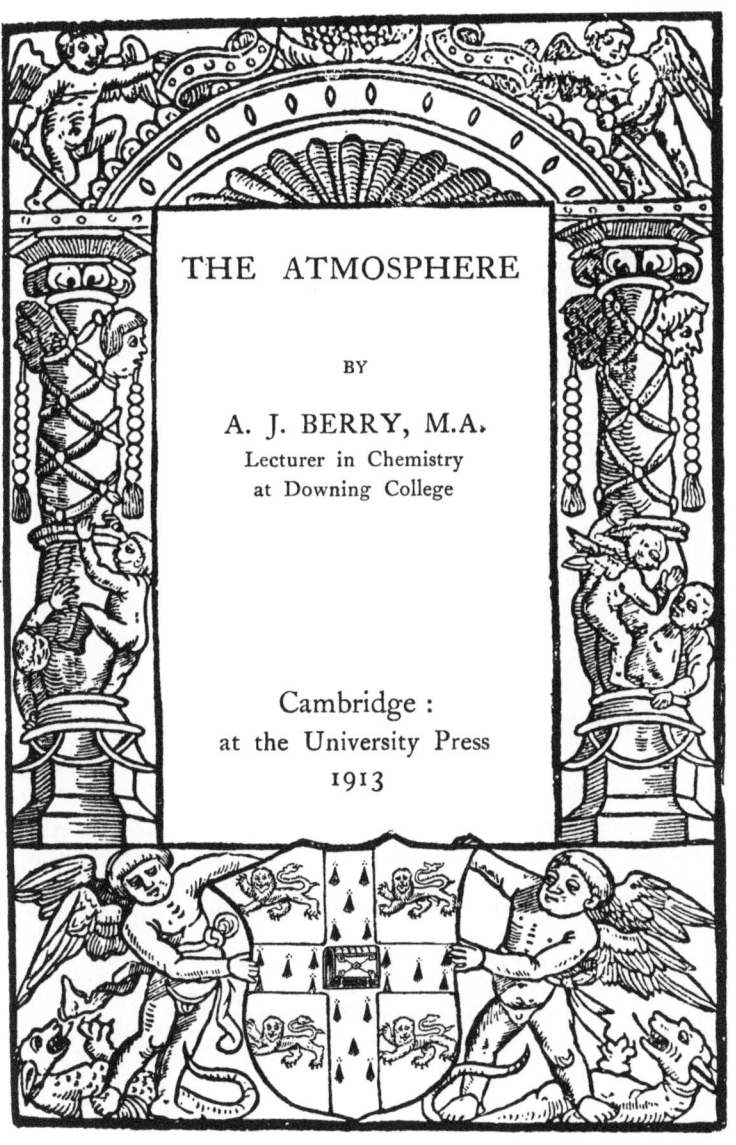

THE ATMOSPHERE

BY

A. J. BERRY, M.A.

Lecturer in Chemistry
at Downing College

Cambridge :
at the University Press
1913

CAMBRIDGE UNIVERSITY PRESS
Cambridge, New York, Melbourne, Madrid, Cape Town,
Singapore, São Paulo, Delhi, Tokyo, Mexico City

Cambridge University Press
The Edinburgh Building, Cambridge CB2 8RU, UK

Published in the United States of America by
Cambridge University Press, New York

www.cambridge.org
Information on this title: www.cambridge.org/9781107401679

First published 1913
First paperback edition 2011

A catalogue record for this publication is available from the British Library

ISBN 978-1-107-40167-9 Paperback

*With the exception of the coat of arms
at the foot, the design on the title page is a
reproduction of one used by the earliest known
Cambridge printer, John Siberch, 1521*

PREFACE

IN the following pages an attempt has been made to give an account of the history of the discovery and of the properties of the constituents of the atmosphere. The subject-matter has been restricted to the more purely chemical and physical phenomena displayed ; meteorology has been omitted.

The author desires to thank Professor Seward for his kind editorial help. To Dr G. F. C. Searle and to the late Mr H. O. Jones he is indebted for many valuable criticisms and suggestions.

A. J. B.

CAMBRIDGE.
November, 1912.

CONTENTS

CHAP.   PAGE

I.   Early history 1

II.   Chemistry during the phlogistic period . . 11

III.   The decline and fall of the phlogistic theory . 21

IV.   The principal constituents of the atmosphere . 33

V.   Modern views on combustion 56

VI.   Constancy of the composition of the atmosphere . 75

VII. The escape of gases from planetary atmospheres
according to the kinetic theory . . . 86

VIII. Liquid air 93

IX.   The inert gases in the atmosphere . . . 103

X.   The radioactivity of the atmosphere . . . 115

XI.   The probable composition of the atmosphere in
early geological time 129

Bibliography 140

Index 143

LIST OF ILLUSTRATIONS

The Hon. Henry Cavendish . . . *Frontispiece*

FIG.   PAGE

The Hon. Robert Boyle 6

Joseph Priestley 15

Joseph Black 17

1. Frontispiece to Priestley's *Observations on different kinds of air* 20

Anton Laurent Lavoisier 23

2. Moody's apparatus 70

3. Dumas and Boussingault's apparatus . . . 80

4. Vacuum vessels 99

5. Rayleigh and Ramsay's apparatus 105

The Frontispiece is from a photograph of the portrait in the Cavendish Laboratory, Cambridge; the portraits of Boyle and Priestley are from those in the National Gallery, London; that of Black is from a Tassie Medallion, by permission of Mr John R. Findlay; fig. 2 is reproduced from the *Journal of the Chemical Society* (1906) by permission of the Society; figs. 3, 4, and 5 are from Roscoe and Schorlemmer's *Treatise on Chemistry* by permission of Messrs Macmillan and Co.

THE ATMOSPHERE

CHAPTER I

EARLY HISTORY

THE discovery of the existence of the atmosphere appears to have been made in very early times. Anaxagoras demonstrated the material nature of the air by its resistance to compression in closed bags of skin. Empedocles stated correctly that air prevents the entry of water into a vessel immersed with its aperture held downwards. But the ideas which people held regarding the nature of the air were exceedingly confused and unsatisfactory. It was not until the seventeenth century that any clear general views were held on the nature of the atmosphere.

In Galileo's time, philosophers explained the phenomenon of suction, the action of syringes and pumps by the so-called *horror vacui*: nature's abhorrence of a vacuum. It must be admitted that apart from the unscientific language in which the idea was expressed, it does to a certain extent really represent the phenomenon. The principle of the 'horror vacui' however was found to break down as it had been found that a pump would not work when the water in the cistern had sunk thirty-five feet below the valve. Galileo does not appear to have had any very clear ideas on the subject. According to one account, Galileo is said to have jokingly remarked that nature's abhorrence of a vacuum did not extend beyond thirty-five feet. It is remarkable that he failed to observe how simply this phenomenon could be explained with reference to the weight of the atmosphere : a fact with which he was well acquainted.

Some time before this, in 1613, Galileo made an attempt to weigh air. He weighed a glass vessel containing nothing but air, and then re-weighed the vessel after the air had been partly expelled by heat. It was known accordingly that the air was heavy. The essential connexion between the phenomena of the 'horror vacui' and the weight of the atmosphere was first clearly demonstrated by Torricelli. Torricelli suggested the use of a column of mercury instead of a column of water to measure the resistance to a vacuum. He expected to obtain a column of about one fourteenth of the length of the water column. His expectation was realised by the now well-known 'Torricellian experiment' which was performed in 1643 by Viviani. A glass tube about one metre in length closed at one end was filled with mercury. The open end was then closed with the finger and inverted in a vertical position in a dish of mercury.

On removing the finger, the mercury forsook the
upper end of the tube and remained stationary at
a height of about 76 centimetres (30 inches).   To
Torricelli belongs the honour of having discovered
the phenomenon of atmospheric pressure.   He also
observed by means of his mercury column that the
pressure of the atmosphere is not invariable but
fluctuates slightly about the mean value of 30 inches.

Torricelli's experiment came to the knowledge of
Pascal in 1644.   Pascal repeated Torricelli's experiment with mercury and also with a tube of water
forty feet in length. The chief merit of this investigator
is to have established a satisfactory analogy between
the phenomena conditioned by atmospheric pressure
and those conditioned by water pressure.   Further re-
flections upon the phenomena of atmospheric pressure
led Pascal to the idea that the barometer column
must necessarily stand lower at the summit of a
mountain than at its base.   This idea was successfully
demonstrated by his brother-in-law, Perier, by an
experiment performed on the summit of the Puy de
Dôme in 1648.

The most important advances in the experimental
study of atmospheric pressure were made by Otto
von Guericke (*Experimenta nova ut vocantur Mag-
deburgica*) about the year 1672.   Guericke seems
to have been a man of extraordinary originality for
he appears to have demonstrated the phenomena of

atmospheric pressure without having been acquainted with Torricelli's experiment. His notions of the nature of the air, its compressibility, and its weight, indicate that he must have been gifted with great powers of observation.

In his first attempt to produce a vacuum, Guericke employed a wooden cask filled with water. He attached the pump of a fire engine to its lower end, and expected after pumping out the water that empty space would remain. However no vacuum was obtained, as the air leaked in through the joints in the cask almost as rapidly as it was pumped out.

After several unsuccessful experiments with wooden barrels, Guericke took a large hollow copper sphere and attempted to exhaust the air directly. At first this was found to be an easy operation, but after a few strokes of the piston, the pumping became exceedingly difficult in consequence of the external pressure, and four strong men could hardly move the piston. At this stage the sphere suddenly collapsed with a loud report. Finally, with the aid of a stronger copper vessel, the production of a vacuum was successfully accomplished.

Guericke also constructed an air pump, and made many interesting experiments with it. He observed the great force with which air rushes into an exhausted space on opening the tap. He also found that a lighted candle was extinguished in a vacuum, and conjectured that it derived its nourishment from the air. He found that sound is not propagated through a vacuum, and that animals die in it.

The most interesting experiments which were made by this investigator however had reference to atmospheric pressure. He constructed an apparatus consisting of two metallic hemispheres, tightly adjusted to one another, which could be exhausted by the air pump. This apparatus has since become known to fame as the Magdeburg hemispheres. When the apparatus was exhausted by the pump, Guericke found that the hemispheres could only be severed by the traction of sixteen horses.

Many other interesting experiments relating to atmospheric pressure were carried out by Guericke, and they created a profound impression on the men of that time. While his investigations were in progress, the phenomena of atmospheric pressure were studied in England by the Honourable Robert Boyle. This remarkable man, who was born in 1626, was the seventh son and fourteenth child of the Earl of Cork, and he made many interesting experiments on atmospheric pressure. In particular, he discovered the law which is known by his name governing the relation between the pressure and volume of a given mass of gas. Boyle also made a number of investigations of considerable chemical importance, to which reference will be made in Chapter II.

The Hon. Robert Boyle (1626—1691).

The law governing the relation between the pressure
and volume of a given mass of gas may be stated in
the following terms :   When the temperature of a
given mass of gas is constant, its volume varies
inversely as the pressure.   Boyle demonstrated this
law by confining a quantity of air by means of
mercury in one limb of a syphon-like tube.   This
shorter limb of the syphon was closed, the longer
limb of the apparatus being open to the external air.
By pouring mercury into the longer limb until the
volume of air in the shorter limb became halved, and
noting the difference in level of the mercury in the
open and closed limbs of the apparatus, he found that
this difference in level was about 30 inches.   In other
words, Boyle showed that by doubling the pressure,
the volume became halved.

Some years after Boyle's work, the same law was
discovered independently by Mariotte.   The investigator discovered the law by partially filling Torricellian
tubes with mercury, measuring the volume of air
remaining, and then performing the Torricellian
experiment.   The new volume was thus obtained,
and by subtracting the height of the new column of
mercury from the barometric height, he also found
the new pressure to which the same quantity of gas
was now subjected.

An important law dealing with the expansion of
gases by heat under constant pressure was discovered

by John Dalton (1766—1844) and also by the French investigator Charles. The law is however usually known as Gay-Lussac's law, since he was the first worker who made experiments having any pretensions towards accuracy to test the validity of the law. The law states that different gases have very nearly the same temperature coefficient of expansion at constant pressure. Gay-Lussac determined the coefficient of expansion of air (and other gases) at constant pressure by confining a known volume of air in a glass bulb connected with a piece of capillary tubing, and observing the expansion of the gas over a given range of temperature by the motion of a small drop of mercury in the capillary. The results obtained by this method were however not very accurate. The first accurate experiments on the coefficient of expansion were made by Regnault.

The law of expansion of gases under constant pressure may be stated in the following terms : The volume of a gas at constant pressure increases or decreases by approximately 1/273 of its volume at 0° Centigrade for every degree Centigrade through which it is heated or cooled.

The laws of Boyle and Charles may be taken as very nearly true over small ranges of temperature and pressure. They are by no means true over very wide ranges of temperature and pressure. These deviations will be considered later.

Torricelli's observation that the height of his mercury column varied slightly from time to time and from place to place has become of practical importance in the construction of barometers, as the instruments for the measurement of atmospheric pressure have been termed. Since Torricelli's time, the mercury barometer has undergone many improvements, the chief improvements being in the direction of affording facilities for reading the height of the mercury column with accuracy.

For accurate work it is also necessary to correct the observed height of the mercury column for the expansion of the mercury and also to allow for the effect on the scale of a rise of temperature. Corrections have also to be applied for the latitude and for the height above the level of the sea, where the indications of the instrument are observed, for certain purposes where a knowledge of the exact value of the atmospheric pressure is required. Standard atmospheric pressure is defined as a pressure equal to that given by a column of mercury 76 centimetres in height, when the temperature of the column is that of melting ice and the instrument is situated at mean sea-level in the latitude of 45 degrees. Since one cubic centimetre of mercury weighs 13·6 grams, the pressure exerted on one square centimetre of surface will be 13·6 × 76 grams or approximately 1033 grams, or nearly 15 pounds on every square inch.

Incidentally, we may refer to the popular use of the barometer as an instrument for the prediction of the state of the weather. As is well known, the barometric height is, generally speaking, greater in fine weather than in unsettled weather, and consequently it is customary to mark the scale of instruments which are intended for weather prediction with certain familiar terms. The prediction of the atmospheric conditions is, however, by no means such a simple matter as is commonly supposed, and observations of the barometric height should in all cases be supplemented with observations of the temperature and of the hygrometric state of the atmosphere, if weather prediction is to be effected with any degree of certainty. It is not the purpose of the present volume to deal with the subject of meteorology, but the reader who desires information on this subject is recommended to consult some special treatise.

A form of barometer which is frequently employed for meteorological purposes is what is known as the aneroid instrument. This instrument records changes in the pressure of the atmosphere by the expansion and contraction of an exhausted metal box. These expansions and contractions are communicated to a system of levers and thereby magnified. The scale of an aneroid barometer has, however, always to be set by comparison with a mercurial instrument. This instrument is frequently employed for obtaining a continuous or graphic record of the changes in atmospheric pressure over a certain interval of time.   In this form it is termed a barograph, and is of great value for meteorological purposes.

CHAPTER II

CHEMISTRY DURING THE PHLOGISTIC PERIOD

THE phenomena of combustion were known for many centuries before any attempt was made to explain them or indeed before any attempt was made to investigate them.   Otto von Guericke's observation, to which reference was made in the previous chapter, that combustion does not take place in a vacuum was of very great importance, but apparently it was absolutely disregarded at the time.

One of Robert Boyle's contemporaries who made some good observations on combustion and calcination was John Mayow (1645–1679).   He assumed that atmospheric air contained a substance which he termed *spiritus igno aereus* or *nitro aereus* which combined with metals to form 'calces' and which was intimately associated with the process of respiration. This assumption contained the germ of the discovery of the now generally accepted explanation of the phenomena of combustion, but it appears to have attracted little attention at the time, and it was not until about one hundred years later that Lavoisier interpreted the facts in a logical and consistent manner.

The theory of combustion which was widely believed during the seventeenth and the greater part of the eighteenth centuries was put forward by Johann Joachim Becher (1635–1682) and especially by Georg Ernst Stahl (1660–1734). According to Becher, all inorganic substances consist of three 'earths,' the mercurial, the vitreous, and the combustible. The last was termed by Becher *terra pinguis*, and combustible substances were supposed to contain the *terra pinguis* in relatively large proportion. The essential part of Becher's hypothesis was that when substances were burnt, the *terra pinguis* escaped.

Stahl elaborated and extended Becher's views on combustion, and he termed the *terra pinguis* which was supposed to be present in such large proportion, phlogiston ($\phi\lambda o\gamma\iota\sigma\tau\acute{o}\nu$). Substances which were readily combustible were supposed to contain a large proportion of phlogiston, while those which only burnt with difficulty were supposed to be relatively poor in that constituent. A metal being calcined was supposed to lose its phlogiston; and when the 'calx' was heated along with coal (a substance very rich in phlogiston) the metal was reproduced.

According to Stahl the relations might be summarised in the following manner :

$$Metal = Calx + Phlogiston.$$

In building up this theory of combustion, no attention was paid to the relations between the weights of the combustible substance and those of its products of combustion. In due course, however, it was discovered that the weights of the 'calces' were greater than the weights of the original metal. The phlogistonists were somewhat puzzled at results of this kind, and attempted to explain the unexpected results by having recourse to the far-fetched assumption that phlogiston possessed a negative weight. Since it was possible to give a plausible explanation of all commonly occurring phenomena by the aid of the phlogistic theory, there is no wonder that it survived for a comparatively long period.

During the phlogistic period, which may be taken to be the period of about one hundred and twenty years after the year 1660, the science of chemistry was enriched by the labours of many eminent investigators, among whom may be mentioned Boyle, Black, Cavendish, Priestley, Scheele, and others. It is most untrue to say, as some French chemists have maintained, that before the time of Lavoisier, chemistry was no science. On the contrary, the discoveries of the phlogistonists were of the highest value.

Guericke knew of only one kind of air. However, the phlogistic period of chemistry must ever be renowned for the discovery of a number of distinct gaseous substances, several of which are constituents of the atmosphere. The discovery of oxygen, or, as it was then termed, 'dephlogisticated air,' was made independently by Priestley in England and by Scheele in Sweden.

There appears to be some misunderstanding as regards the relative claims of these two investigators to the priority of the discovery of 'dephlogisticated air.' The truth appears to be that Scheele actually discovered oxygen and made a study of its properties between the years 1771 and 1773, while Priestley first isolated the gas on August 1st, 1774. The latter investigator, however, preceded Scheele in the publication of his work, and is therefore frequently regarded as the original discoverer of oxygen. But there can be little doubt that the actual isolation of the gas was first effected by the Swedish chemist.

Both Priestley and Scheele observed that the gas supported combustion and respiration in an intensified degree. The gas was given the names of 'fire air' and 'life air,' but the name which found most favour was 'dephlogisticated air.'

Priestley, who has been named the Father of pneumatic chemistry, discovered a number of distinct gaseous substances. His success in this sphere was

Joseph Priestley (1733—1804).

truly remarkable.   He introduced the method of collecting over mercury those gases which are readily soluble in water.   In this way he isolated the following gases : ammonia, hydrogen chloride, silicon fluoride, and sulphur dioxide.   The gas known as 'inflammable air' (hydrogen) was discovered by Cavendish.

The question which above all others attracted the attention of the chemists of that time was the question of the nature of atmospheric air.   This question was solved experimentally by the labours of several chemists of the phlogistic period, although the correct interpretation of their results was not given until the overthrow of the phlogistic theory by Lavoisier.

The germ of the discovery that atmospheric air is not a simple substance, but contains a constituent capable of supporting combustion and respiration was made by Boyle and by Mayow, although neither of these investigators succeeded in isolating that constituent.   The actual isolation of dephlogisticated air was not made until about one hundred years later.

The discovery of the gas which was known in the phlogistic period as 'fixed air' was made many years previously by Van Helmont, an iatro-chemist of the school of Paracelsus.   But Van Helmont's observations were forgotten, until the Scotch chemist Joseph Black rediscovered the gas in 1752.   Black's discovery of 'fixed air' had its origin in a masterly attempt to elucidate the differences between the so-called 'mild' and 'caustic.' alkalies. He demonstrated the acidic nature of 'fixed air' for it converts quicklime, the acid earth as he termed it, into crude lime or mild earth, the mildness being due to the union with 'fixed air.' Black was an admirable experimentalist: his work was characterised by precision and tolerable

Joseph Black (1728—1799).

accuracy. He was the first chemist to weigh a gas by observing the increase of weight of a substance after absorption of the gas.

The discovery of nitrogen, or as it was variously termed 'mephitic air' and 'phlogisticated air,' was made independently by Rutherford and Scheele. Rutherford's research was, as he himself acknowledged,

inspired by Black's experiments on 'fixed air.'
Rutherford showed that when oxygen ('dephlogisticated air') was removed from ordinary atmospheric air
by combustion or by respiration, and the resulting
'fixed air' absorbed by alkalies, the residual gas was incapable of supporting combustion or respiration. The
residue became consequently known as nitrogen, and
although incapable of supporting combustion, was
first termed 'phlogisticated air.'

Notwithstanding all the momentous discoveries
made during the phlogistic period, a correct explanation of the phenomena of combustion was not attained.
As a matter of fact, the chemists of that period were
thoroughly fettered by their belief in phlogistonism,
and, unfortunately, they attempted to defend the theory
against the experimental facts which they themselves
discovered. The most brilliant investigator of that
time, the Honourable Henry Cavendish (1731–1810),
whose works contributed more than those of any
other experimentalist towards the overthrow of the
phlogistic theory, remained a firm adherent to phlogistonism. Cavendish discovered many things which
are commonly associated with the names of subsequent
investigators, and it was only when his manuscripts
were edited by the late Professor Clerk Maxwell that
it was seen that he was far in advance of the science
of his time. Cavendish made important contributions
to almost every branch of natural knowledge ; it is

however with his more purely chemical work that we are concerned here.

It had been known for centuries that a gas could be evolved by the action of certain metals on some acids. The exact nature of this gas remained unknown until Cavendish investigated its properties about the year 1766. Cavendish found that the gas was readily inflammable. He regarded it as a compound of phlogiston and termed it 'inflammable air.' The further discovery of the compound nature of water, which was also made by Cavendish and which proved to be the death-blow to the phlogistic theory, will be considered in the next chapter.

The most serious difficulty with which the phlogistic theory had to contend was the fact that the products of combustion possessed a greater weight than the combustible substances from which they were derived. Attempts were made again and again to show that air must be present in order to take the place of the escaping phlogiston. The ingenious attempt to explain the phenomena by assuming that phlogiston possessed a negative weight, and that therefore the weight of the products of combustion *must* be greater than the weight of the combustible material was regarded by many as plausible, although it must be admitted that there were a few chemists who protested against such an explanation. It remained for Lavoisier to interpret the facts correctly.

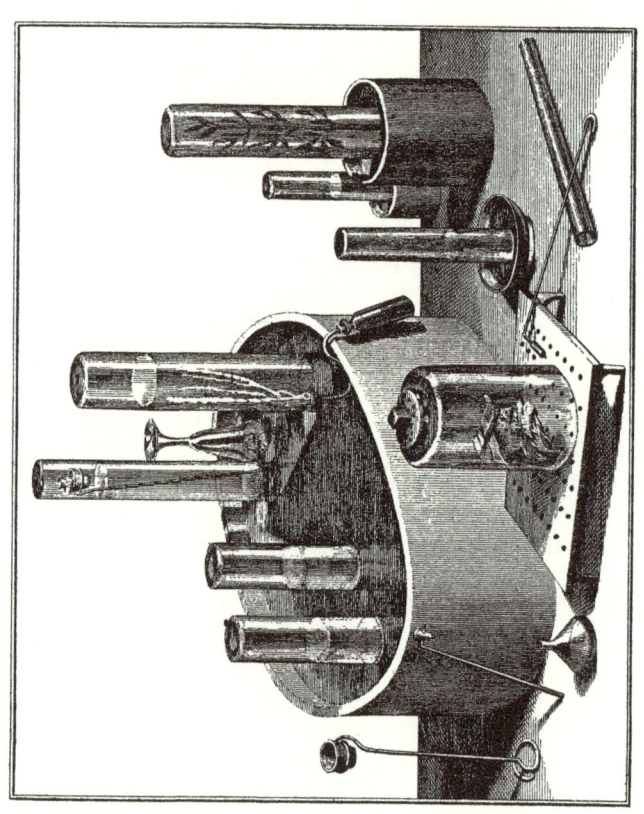

Fig. 1.   Frontispiece to Priestley's *Observations on Different Kinds of Air.*

Important progress was made in other branches of chemistry during the phlogistic period. A number of new elements were discovered ; among these may be mentioned phosphorus, chlorine, cobalt, manganese, and platinum. The discovery of phosphorus by Brand and by Kunkel was one of the most important discoveries of the period. Technical chemistry too made enormous strides and many organic compounds were prepared, especially by Scheele. Before the time of Boyle chemistry was simply the handmaid of medicine, but during the phlogistic period it assumed the character of an independent science. Chemistry also entered upon a new phase by the opening up of common ground with other closely related sciences, physics on the one hand and biology on the other.

CHAPTER III

THE DECLINE AND FALL OF THE PHLOGISTIC THEORY

WE have seen in the preceding chapter that the one insuperable difficulty in the way of the phlogistic theory becoming universally accepted was its failure to explain the fact that the products of combustion of a combustible substance possess a greater weight than the substance itself. The overthrow of the phlogistic theory was the work of the distinguished French investigator Anton Laurent Lavoisier (1743–1794). Lavoisier was essentially a physicist, his interest in chemical phenomena was directed more particularly to their quantitative relations. It is therefore not surprising that he laid great stress on the alterations in weight resulting in the combustion of various substances ; the properties of the products were of comparatively little interest to him. Lavoisier was fortunate in having a valuable collection of experimental data at his disposal which had been accumulated by the phlogistic chemists. Unfortunately, although he made free use of the discoveries of others, in particular of Priestley's discovery of oxygen (dephlogisticated air), of Black's discovery of carbon dioxide (fixed air), and especially of Cavendish's discovery of the compound nature of water, he was most unscrupulous in wilfully refusing to acknowledge the great assistance which he derived from those discoveries.

Lavoisier commenced his researches on combustion in the year 1772. In a sealed note delivered to the Académie des Sciences on the 1st of November of that year, he communicated the results of some experiments which demonstrated that by the combustion of sulphur and phosphorus, the weights of the volatile products of combustion are greater than those of the combustible substances. It was also shown that by reducing litharge (lead monoxide) by

Anton Laurent Lavoisier (1743—1794).

heating it with coal in a closed vessel, a considerable volume of gas was evolved—about one thousand times the volume of the litharge. It is important to observe here that at this stage Lavoisier really made very little advance on Mayow, who had observed about one hundred years previously that substances on combustion give rise to products of greater weight. At this stage of his work Lavoisier was quite ignorant as to which constituent of the air was concerned with the increase in weight.

The circumstances in which Lavoisier became acquainted with the discovery of 'dephlogisticated air' are remarkable. Priestley being in Paris with Lord Shelburne at the time was dining with Lavoisier, and incidentally mentioned his discovery of 'dephlogisticated air.' He stated that he had prepared the gas by heating mercuric oxide (*precipitate per se*), and also by heating red lead. The gas obtained was found to be a powerful supporter of combustion. Lavoisier soon repeated Priestley's experiments and confirmed his results.

In 1774 Lavoisier communicated some further experiments to the Académie des Sciences. These experiments referred to the calcination of tin in sealed retorts. It was shown that when a quantity of tin was heated in a sealed retort, the weight of the apparatus and its contents was the same before and after being heated. On opening the retort, however, air rushed in, and the apparatus and its contents increased in weight. This increase in weight was found to be equal to that which the same weight of tin had undergone on calcination with free access of air. It was therefore concluded that the calx of tin is a compound of tin and air. Between 1774 and 1777 Lavoisier, having become familiar with the discovery and properties of oxygen, and also with Black's work on carbon dioxide, was steadily classifying facts for the antiphlogistic theory of combustion. Among other interesting observations, he discovered that the diamond on combustion yields carbon dioxide, and that in many cases the properties of the products possess acidic characters. In the year 1777 Lavoisier gave his theory of combustion to the world, the chief points of which are as follows :

(1) Substances only burn with free access of air.

(2) The air is consumed during the process of combustion, and the increase in weight of the substance burnt is equal to the decrease in weight of the air.

(3) The combustible substance is frequently converted into an acid, but the metals on the other hand into metallic calces.

This third statement contains the essence of Lavoisier's theory of the composition of acids. He believed that all acids contain oxygen, and that oxygen is *par excellence* a substance which imparts acidic properties. For some time Lavoisier was occupied with the problem as to the nature of the acid produced by the combustion of hydrogen. This problem was solved, not by Lavoisier but by the phlogistonist Cavendish, who showed that water is the sole product of the combustion of hydrogen.

The discovery of the compound nature of water has claimed the attention of many who have written on the history of chemistry. The facts of the case appear to be briefly as follows. In 1781 Priestley performed what he termed 'a random experiment' merely for the purpose of entertaining a few philosophical friends. This experiment consisted in exploding a mixture of 'inflammable air' and common air contained in a closed vessel by the aid of the electric spark in the manner first practised by Volta in 1776. Priestley's friend, John Warltire, who witnessed the experiment, called Priestley's attention to the fact that after the explosion the sides of the vessel were covered with a film of moisture. The matter was next investigated by Cavendish with characteristic thoroughness. He found that the greatest diminution of volume occurred on sparking a mixture of two volumes of 'inflammable air' and five volumes of common air, and he found that moisture was produced as in Priestley's experiment. This moisture was examined carefully and was found to be identical in properties with pure water. Cavendish recognised quite clearly that only one portion of atmospheric air was concerned in the formation of water, and that this portion was the 'dephlogisticated air.' In his next experiment, he subjected a mixture of two volumes of 'inflammable air' to one volume of 'dephlogisticated air' to the action of the electric spark in a special apparatus which he termed a eudiometer. Water was formed as in the previous experiment. Cavendish however delayed the publication of his results for some time, with the consequence that the question of priority in this discovery gave rise to an embittered controversy afterwards.

In 1783 Lavoisier was informed verbally by Blagden of Cavendish's discovery which was published in 1784. Lavoisier repeated Cavendish's experiments on the combustion of hydrogen in oxygen and obtained a certain quantity of water. There appears to be no doubt, however, that Lavoisier would never have arrived at Cavendish's conclusion had he not been informed of the latter's discovery by Blagden. In connexion with the discovery of the compound nature of water, mention must be made of the contribution of James Watt. This distinguished engineer was profoundly interested in Priestley's experiments, and inaccurate as these experiments were, he appears to have been the first to realise that water is not an elementary substance. However, for the experimental proof of the quantitative composition of water the merit is undisputedly due to Cavendish.

Lavoisier's reasoning in 1777, and the subsequent discovery of the compound nature of water, sealed the doom of the phlogistic theory. It is perhaps somewhat remarkable that Priestley and Cavendish were never converted to the new views. Of all the chemists of the phlogistic period, one only, Black, saw the great advantages of the Lavoisierian theory of combustion, and released himself from the fetters of phlogistonism. It is easy for us in the twentieth century to ridicule the absurdities of the phlogistic theory, but if it be borne in mind that the idea of phlogiston is very closely allied to energy, the theory was undoubtedly of value, inasmuch as it served to stimulate a vast amount of experimental research.

The downfall of the phlogistic theory may be said to be the beginning of a new era in chemistry: it is in truth the commencement of modern chemistry. But the phlogistic period will always be associated with the discovery of the more important constituents of the atmosphere, and after the fall of phlogistonism the attention of chemists was directed into other paths. It is not the purpose of this volume to attempt to write a general history of chemistry after the phlogistic period, but some reference must be made to the development of the conceptions of atom and molecule in the science.

The commencement of the nineteenth century will ever remain a memorable epoch in the history of chemistry.   Between the years 1801 and 1808 a keen controversy raged between two French philosophers, Berthollet and Proust, regarding the question as to whether chemical combination takes place between definite proportions of the reacting substances or whether the composition of compounds depends upon the manner in which they are produced.   The careful experimental work of Proust demonstrated the fact that the composition of compounds is perfectly definite.

However the fundamental laws of chemical combination were first clearly stated by John Dalton. This investigator revived a theory of certain Greek philosophers who taught that matter is made up of minute indivisible particles termed atoms.   Dalton, however, was the first to give the atomic theory an essentially chemical significance, in particular he laid great stress on the conception of mass. According to Dalton the atoms of the different elements are not identical in weight, and the relative atomic weights of the elements are the gravimetric ratios in which these elements combine together or are multiples of these.

This theory, which was propounded in 1803, was probably arrived at by Dalton deductively, and not as the result of his experimental work on chemical combination.   His experiments on the combination of gases led to certain volumetric relations which necessitated an amplification of the atomic theory. Dalton made the assumption that equal volumes of gases, under similar physical conditions, contain the same number of atoms, but he found that it was necessary to reject this assumption as being inconsistent with the facts. The correlation of the volumetric relations of gases and the atomic theory was made in 1811 by the Italian physicist Amadeo Avogadro, who framed the hypothesis that equal volumes of gases, under the same conditions of temperature and pressure, contain the same number of molecules. Avogadro thus introduced the conception of the molecule into chemistry. A molecule may be defined as the smallest part of an element which can exist in the free state, while an atom is defined as the smallest part of an element which is capable of entering into combination. It is evident from what has been stated that Dalton made no distinction between the ultimate particle of an element and the ultimate particle of a compound: both were alike termed atoms. The necessity of making such a distinction was first clearly recognised by Avogadro.

Dalton drew up a table of relative atomic weights which he calculated from the results of his own experiments. However, the first really accurate atomic weight determinations were made by the Swedish chemist Berzelius. It is remarkable that in a number of cases, the values of the atomic weights in use at the present time do not differ greatly from those of this admirable experimentalist.

In the course of his experimental work, Dalton discovered the law usually known as the law of multiple proportions, according to which the same element may combine with another element in varying simple multiple proportions. At the present time this law has almost entirely lost its significance ; the innumerable compounds of the element carbon afford little evidence of combination by simple multiple proportion.

It has been necessary to make this slight digression, as there is scarcely a field of investigation in modern chemistry which is not concerned with atoms and molecules. In this connexion it is interesting to note that recent investigations of certain radioactive phenomena have proved the physical reality[1] of atoms and molecules. This is a result of the highest importance : until quite recently the conceptions of atom and molecule were regarded as convenient assumptions which were employed to explain or rather to describe certain phenomena, whereas we now know that these terms connote definite physical realities.

[1] The discovery of the property of asymmetry among certain organic compounds gave very strong evidence in favour of the physical reality of atoms and molecules. Support to the same idea has also been obtained by a study of the so-called ' Brownian movement.'

An important relation connecting the specific heat of an element with its atomic weight was discovered in 1819 by Dulong and Petit. These investigators enunciated the law that the product of the specific heat of an element in the solid state and its atomic weight is approximately constant, or in other words that the atoms of elements have all the same thermal capacity. This law has been of value in controlling doubtful atomic weights where an element was found to combine with other elements in more than one proportion. This law is however purely empirical, it has no theoretical foundation whatever.

The most remarkable relation between the atomic weights of the elements and their properties was pointed out in 1869 by Mendeleeff and also by Lothar Meyer. Many years previously, attempts had been made to establish relations between the values of the atomic weights of certain elements and their properties, notably by Döbereiner and by Gmelin. In 1864 a somewhat premature announcement was made by Newlands, who stated his now well-known 'law of octaves.' But the honour of having elaborated the periodic law, and of having predicted the existence of unknown elements by its aid, belongs to the Russian chemist Mendeleeff.

CHAPTER IV

THE PRINCIPAL CONSTITUENTS OF THE ATMOSPHERE

WE have seen that we are indebted to the labours of the phlogistic chemists for the discovery of the chief constituents of the atmosphere, and to Lavoisier for the correct interpretation of the phenomena of combustion and oxidation. It is the purpose of the present chapter to give an account of the more important properties of these gases, and of the part which they play in the economy of nature.

Nitrogen.

Nitrogen was found by the older chemists to be chiefly characterised by its inertness. It is a colourless gas which is very slightly soluble in water, and is incapable of supporting combustion. It can however be caused to unite with oxgyen by the prolonged action of electric sparks or of the electric arc : in such circumstances it forms oxides of nitrogen. At the present time this union of nitrogen and oxygen under the influence of the arc has become of technical importance as it is being employed for the manufacture of nitric acid. Nitrogen is also capable of uniting with certain metals at moderately high temperatures to form nitrides. This gas is also a constituent of a vast number of organic compounds, and many interesting inorganic compounds are derivatives of this somewhat inactive element. But, generally speaking, numerous as the derivatives of this gas are, they are usually only obtainable by indirect methods.

It would be out of place in a book of the present kind to attempt to give an account of the numerous methods of preparing the gas and its more important compounds. For such, the reader is referred to a suitable text-book. As regards the preparation of the gas, it is sufficient to state that there are two general methods of procedure. The first general method is to prepare the gas from the atmosphere by the removal of the other constituents by the use of suitable reagents. The other general method is to prepare the gas from some one or other of its numerous compounds. While the latter method is capable of yielding a product of absolute purity, the product of the former is invariably contaminated with certain rare gases which are present in the atmosphere. (See Chapter IX.)

For the preparation of atmospheric nitrogen, the usual procedure is to absorb the carbon dioxide by means of caustic potash, and the aqueous vapour by means of some desiccating substance such as calcium chloride or concentrated sulphuric acid, and then to remove the oxygen by means of some more or less oxidisable substance such as phosphorus or preferably by some metal such as copper. The product obtained in this manner consists under favourable conditions of 99 per cent. nitrogen, the inert gases constituting the one per cent. impurity.

The physical properties and constants of nitrogen have been the subject of careful investigation by various observers. The gas has been liquefied and solidified by the application of extreme cold, and the physical constants of the liquid and solid have been determined in recent years. We shall return to this subject in Chapter VIII.

It has already been stated that elementary nitrogen is characterised by relatively great chemical inertness. But numerous instances are known where an element exhibits the phenomenon of allotropy, which may be defined as the property of an element existing in two or more modifications, the various modifications possessing characteristic properties, and in 1911 the Hon. R. J. Strutt showed that nitrogen can be converted into an allotropic modification. This investigator found that when nitrogen at low pressures is subjected to the action of the electric discharge from a Leyden jar, the gas continues to glow after the discharge has ceased. With the object of examining the nature of this after-glow, a current of nitrogen was drawn through the discharge tube into an observing vessel, the current being maintained by a mechanical pump, and the after-glow subjected to spectroscopic investigation. It was found that the glowing nitrogen possessed a characteristic band spectrum. The glow is extinguished by the action of heat, and intensified by cooling the tube, although it is completely extinguished at very low temperatures. The action of the modified nitrogen on various metals and metalloids was studied, and it was shown that the gas was possessed of very considerable chemical activity. The proportion of the total nitrogen which becomes 'activated' by the oscillatory discharge is relatively small; Strutt estimates it at about $2 \cdot 5$ per cent. The field of investigation which has here been opened out promises to be one of surpassing interest, and much experimental work remains to be done before the precise nature of the active nitrogen becomes established.

The function of the nitrogen in the atmosphere is largely that of a diluent of the oxgyen. This element however plays a most important part in the life of plants and no account of the properties of the gas can possibly be regarded as complete without some reference, however inadequate, to the functions which it performs in the metabolism of the plant. Most plants are incapable of assimilating nitrogen directly from the atmosphere, but they derive their supply of this element largely from the soil. There are however some plants, notably members of the Leguminosae, which do derive their nitrogenous supply from the atmosphere. These plants possess peculiar tubercles on their roots which are filled with a bacterial mass consisting partly of hypertrophied bacterioids and partly of normal bacteria. It has been shown by the work of Beijerinck, Hellriegel and Wilfarth, and others that we have here to deal with a case of symbiosis. While the bacteria (*Bacillus radicicola*) flourish on the carbohydrates supplied by the host plant, the latter derives its nitrogenous supply from the atmospheric nitrogen which is fixed by the bacteria. It had been known for centuries that the Leguminosae were capable of flourishing on soil poor in nitrogen, and of enriching the soil by this accumulation of nitrogenous organic matter, but the essentially symbiotic nature of the phenomenon has only been understood in comparatively recent years. Assimilation of nitrogen by root tubercles has also been observed in some genera which are not members of the Leguminosae.

The conversion of nitrogenous matter in the soil into nitrates is a process which is effected by two organisms. One of these organisms is concerned with the oxidation of the ammonia and protein matter to nitrites, while the other effects the oxidation of the nitrites to nitrates. Nitrification can only proceed when some basic substance is present in the soil which

neutralises the free acid produced, and is dependent
on certain conditions favourable to the life of the
organisms. The converse process, viz., the conversion
of nitrates into elementary nitrogen, is continually
taking place to a considerable extent on the surface
of the earth under the influence of the so-called
denitrifying bacteria, and the combined nitrogen is
thereby reconverted into the gaseous form.

The Technical Fixation of Atmospheric Nitrogen.

At the present time, the principal sources of nitrogen compounds are (1) Chili saltpetre (sodium nitrate)
from which the bulk of the nitric acid of commerce
is obtained, and (2) ammonia obtained as a by-product
in indust ries where coal is carbonised. But the
supplies of Chili saltpetre and of coal are not expected
to last indefinitely, and processes have been devised
for the fixation of nitrogen on a technical scale. Two
methods of effecting the fixation have been devised.
The first depends on the union of nitrogen and oxygen
in the electric arc resulting in the formation of oxides
of nitrogen, which are absorbed by potash with formation of potassium nitrite and nitrate ; in some cases
lime water is employed to absorb the oxides of
nitrogen, so that the nitrogen is obtained in the form
of calcium nitrate. The second method depends on
the union of nitrogen with calcium carbide at the

high temperature of an electric furnace resulting in
the formation of calcium cyanamide :

$$CaC_2 + N_2 = CaCN_2 + C.$$

This substance readily yields ammonia by the action
of water,

$$CaCN_2 + 3H_2O = CaCO_3 + 2NH_3,$$

and on this account is of great value as a fertiliser.

Several processes have been devised for the manufacture of nitrates by the first method.   Bradley and
Lovejoy in 1902 patented a process which was worked
at the Niagara Falls in which air previously cooled and
dried was subjected to the action of a series of rapidly
interrupted arcs. This process, however, proved to be
a failure owing to the wear and tear of the plant and
to the poor yield of the product.   A very successful
process, patented by Birkeland and Eyde, is being
worked at Notodden in Norway, where water-power
of great extent is available.   In this process, the arc
is spread out by the action of powerful electromagnets
and the oxides of nitrogen are absorbed by water
until nitric acid of 50 per cent. strength is obtained.
The gases are then passed into milk of lime, and also
over dry lime, whereby calcium nitrate is chiefly
formed.   By subsequent treatment with nitric acid,
a crude calcium nitrate is obtained which can be used
directly as a fertiliser of the soil.   The hygroscopic
nature of this product is overcome by converting it

into a basic nitrate by mixing it with lime and calcining
the product.

One of the most important constants of an element
is its atomic weight, and while it is not intended to
enter into details regarding the methods of atomic
weight determination, it is necessary to state that
this gas possesses a diatomic molecule, and that the
atomic weight of this element at present adopted is
14·01 in terms of oxygen as 16·000.

Oxygen.

The second constituent of the atmosphere, viz.
oxygen, is an element of very great importance.
Although present in the atmosphere in approximately
one-fifth part by volume as compared with four-fifths
of nitrogen, the fact that this gas is concerned with
the phenomena of combustion and respiration at once
claims our attention. In the combined state, oxygen is
a constituent of water and of a very great number of
inorganic and organic compounds. Oxygen is a
colourless, tasteless, and odourless gas, which is
possessed of very considerable chemical activity. It
is more soluble in water than nitrogen. The chief
interest of this gas is concerned with the phenomena
of combustion ; but on account of the very great
importance of these phenomena, it is necessary to
devote a special chapter to them (Chapter V).

For experimental purposes, the gas may be prepared in many ways. It may be obtained by heating numerous compounds which contain the element. It may also be obtained by the fractional distillation of liquid air ; this method depends upon the fact that liquid nitrogen possesses a lower boiling-point than liquid oxygen, and an easy separation may be thereby effected. It may also be obtained by the electrolysis of certain aqueous solutions and in numerous other ways, but it would serve no useful purpose to describe them. Many beautiful experiments can be performed by burning substances in the pure gas, and in the liquid and solid states the element displays very interesting properties. Sir James Dewar has shown that liquid oxygen is so strongly paramagnetic as to rush in drops to the poles of a powerful magnet.

The simplest derivatives of oxygen are termed oxides, and are of two general types. Non-metallic oxides are possessed of acidic properties in presence of water, whereas the oxides of metals are of a basic character. Most elements are capable of forming more than one oxide. There is also a vast number of compounds containing two or more elements besides oxygen.

Inasmuch as oxygen is possessed of great chemical activity and as its compounds are well defined, and easily prepared in a high degree of purity suitable

for quantitative analysis, this element has been adopted
as the standard for atomic weight determinations.
Dalton selected hydrogen as his standard for atomic
weights, as it is the lightest known gas. Berzelius
selected oxygen, since hydrogen forms comparatively
few compounds which are readily capable of accurate
analysis. Careful experiments have shown that the
atomic weights of these two gases are approximately
in the ratio of 1 to 16. In recent years it has been
shown, as the result of the work of Lord Rayleigh,
Leduc, Morley, Scott, and others, that the ratio is not
exactly 1 to 16, but approximately 1 to 15·88, or
1·008 to 16. The latter method of expressing the
ratio is the one which has been adopted generally
at the present time, all atomic weights being expressed
in terms of oxygen as 16·000.

The allotropy of oxygen has been known for
many years. In 1785 Van Marum found that oxygen
underwent a change of properties on exposure to the
electric spark, but the observation attracted no
attention at the time. In 1840 the phenomena were
re-investigated by Schönbein. This chemist found
that the modified oxygen was capable of effecting
a number of oxidising actions which are not effected
by ordinary oxygen. Later, it was shown that oxygen
was capable of becoming 'activated' in many other
ways. The active oxygen was termed ozone by
Schönbein on account of its powerful odour. Ozone

appears to be formed in certain cases where readily
oxidisable substances are allowed to undergo slow
oxidation. These phenomena will be considered in
the next chapter.

For many years much doubt existed as to the
nature of the active oxygen, but in 1856 Andrews
showed that the product possesses identical properties
whatever be the source from which it is derived, and
also that it is simply allotropic oxygen, not a compound
of oxygen with some other element.

In ordinary circumstances, it is not possible to
transform more than a small percentage of oxygen
into ozone by the silent discharge, the best conditions
being to maintain the gas at a moderately low temperature, and to employ a fairly high tension discharge.
The transformation of oxgyen into ozone is what is
known as an endothermic process, a large quantity
of heat being absorbed. When ozone is heated, however, it is reconverted into ordinary oxygen. How
then is this apparently contradictory phenomenon
to be reconciled with the fact that ozone is an endothermic substance? The explanation in brief is as
follows. At every temperature, there is a definite
equilibrium between oxygen and ozone, rise of temperature tending to transform the oxygen into ozone,
lowering of temperature favouring the reverse reaction.
But the velocity of all chemical reactions is profoundly
affected by change of temperature ; as a general rule,

a rise of temperature of 10° C. doubles or trebles the reaction velocity. Further, the decomposition of ozone is a reaction which at moderate temperatures takes place with much greater velocity than its formation from oxygen. It is necessary to distinguish carefully between the two effects of raising the temperature, firstly, its effect on chemical equilibrium and, secondly, its effect on the reaction velocity.

The gas, which has a faint bluish colour, possesses an unpleasant odour and very powerful oxidising properties. Metals such as silver and mercury which are entirely unaffected by oxygen at ordinary temperatures are at once oxidised by ozone. This allotrope of oxygen also combines with certain organic compounds forming peculiar derivatives termed ozonides. At very low temperatures it may be condensed to a very deep blue, strongly magnetic, and exceedingly explosive liquid.

The earlier investigators observed that the ozonisation of oxygen was accompanied by a diminution of volume, and this fact gave the clue to the elucidation of the nature of ozone. In order to follow the reasoning by which the nature of the molecular condition of ozone has been established, it is necessary to make a slight digression.

When hydrogen and oxygen in the proportions of two volumes of the former to one of the latter are caused to unite under the influence of the electric spark, a contraction ensues, the volume of the gaseous product being reduced to two-thirds of the volume of the mixed gases, provided the steam be maintained in the gaseous condition. In other words, two volumes of hydrogen and one volume of oxygen yield two volumes of steam. According to Avogadro's hypothesis, two molecules of hydrogen and one molecule of oxygen yield two molecules of steam. If the molecules of hydrogen and oxygen consist of one atom only, we should reason that two molecules of steam are the product of two atoms of hydrogen and one atom of oxygen, or in other words that one molecule of steam is composed of one atom of hydrogen and half an atom of oxygen. Since atoms are supposed to be indivisible, it is necessary to assume that the molecules of hydrogen and oxygen are composed of two atoms. The formation of water is thus expressed by the equation :

$$2H_2 + O_2 = 2H_2O.$$

Now in the conversion of oxygen into ozone, it has been found that three volumes of oxygen condense to form two volumes of ozone. Since oxygen is a diatomic gas, it is necessary to assume that the molecule of ozone consists of three atoms.

The fact that the molecule of ozone is triatomic was proved in a totally different manner by Soret in 1868. Graham in 1846 discovered the laws governing the rates of diffusion of gases. He showed that the relative rates of diffusion of different gases are inversely proportional to the square roots of their densities. By comparing the relative rates of diffusion of ozone and chlorine, a gas which is known to possess a diatomic molecule, Soret found that the density of ozone is one and a half times as great as that of oxygen. Since the density of a gas is one half of its molecular weight, it follows that the molecule of ozone consists of three atoms.

It has been stated from time to time that ozone exists in the atmosphere especially near the seaside, but considerable doubt exists as to the truth of the statement. The tests which have been employed hitherto to demonstrate the presence of ozone are inconclusive, inasmuch as positive results are also obtained by other substances such as the higher oxides of nitrogen which are undoubtedly formed in the atmosphere by electric discharges. It is, however, quite possible that small quantities of ozone may be formed during a thunderstorm.

Carbon Dioxide.

The third important constituent of the atmosphere, carbon dioxide, or as it is variously termed carbonic anhydride or carbonic acid, although present in minute proportion (approximately 3 volumes to 10,000 volumes of air) is of very great importance. Reference has

already been made to Black's investigations on this
gas. We are indebted to Lavoisier for having demonstrated the fact that this gas is produced by the
combustion of carbon in oxygen. When carbon is
burnt in oxygen the volume of the carbon dioxide
formed is equal to the volume of the oxygen which is
consumed. This fact establishes the composition of
the gas.

For experimental purposes the gas may be prepared from many metallic carbonates, either by
dissociating the carbonates by raising their temperatures, or by decomposing them with a dilute acid.
Carbon dioxide is a colourless gas which is moderately
easily liquefied and solidified. It is moderately soluble
in water ; at the ordinary temperature, water absorbs
approximately about an equal volume of the gas, and
it is incapable of supporting combustion and respiration. This gas, however, plays a most important part
in the organic world, inasmuch as it is the source from
which green plants, i.e. those which contain chlorophyll,
derive their carbon. This process, usually known
as carbon assimilation or photo-synthesis, has received
a great deal of attention from investigators both from
the botanical and the chemical standpoint.   The conditions under which photo-synthesis takes place are,
firstly, the presence of sunlight, of which the red rays
of the spectrum are the most effective, secondly, the
presence of the chlorophyll granules, and thirdly, a

suitable supply of water containing certain salts in solution. By a marvellous process, the precise nature of which is very imperfectly understood, the carbon dioxide of the atmosphere is transformed into the carbohydrates which are found in the body of the plant, and at the same time, a quantity of oxygen is liberated which is very nearly equal to that of the carbon dioxide which is assimilated. Numerous chemists have attempted to follow the intermediate stages of the process. In 1870 Baeyer suggested that a substance known as formaldehyde (CH_2O) might possibly be one of the first stages of the reduction of the carbon dioxide. The preparation of carbohydrates from carbon dioxide thus became a possibility as it was known that carbon dioxide could be converted into formaldehyde by more or less drastic processes. In 1907 it was shown by Dr Fenton that the reduction of carbon dioxide to formaldehyde could be carried out in aqueous solution at ordinary temperatures, so that it is now possible to effect the complete synthesis of sugars under conditions such as prevail in the plant ; and in 1912 it was proved beyond doubt by Curtius and Franzen that formaldehyde is actually produced during photosynthesis.

The carbon dioxide in the atmosphere is derived from a variety of sources. It is formed by the respiration of animals and plants as well as by the combustion of organic matter. It is also discharged into the atmosphere in relatively considerable quantities from volcanoes and from fissures in the earth. One of the most notable of the latter is the Grotto del Cane near Naples which discharges the gas in considerable quantities so that small animals when thrown into the cave are speedily suffocated. The gas is also formed by the decomposition of organic matter in the soil from which it is absorbed by rain and spring water. The refreshing taste of the latter is largely due to dissolved carbon dioxide.

It appears to be probable that the quantity of carbon dioxide in the atmosphere has varied considerably during different periods of geological time. It has been contended that during the Carboniferous period, the percentage of this gas in the atmosphere was greater than at the present time. By the growth of a luxuriant vegetation the excess of carbon dioxide became removed from the atmosphere and the organic matter of the plants gradually became converted into coal. This subject will be considered in a subsequent chapter.

At the present time it would appear that the approximate constancy of the carbon dioxide content of the atmosphere is due to the combustion of carbonaceous matter on the one hand, and to the assimilation of the gas by green plants on the other. Another most important regulator of the carbon dioxide content of the atmosphere is the sea, as Peligot

observed in 1855. In 1880 Schloesing pointed out
that considerable quantities of carbon dioxide are
absorbed by sea water with formation of the soluble
calcium bicarbonate. In the sea there is a definite
equilibrium between calcium bicarbonate, normal
calcium carbonate, carbon dioxide, and water, as indicated by the equation :

$$Ca(HCO_3)_2 \rightleftarrows CaCO_3 + H_2O + CO_2.$$

If from any cause the proportion of carbon dioxide
in the atmosphere becomes excessive, this excess of
gas dissolves in sea water resulting in the formation
of fresh calcium bicarbonate. But if the atmospheric
carbon dioxide falls below the normal amount, the
sea gives up some of its dissolved carbon dioxide,
and fresh calcium bicarbonate decomposes into the
normal carbonate, carbon dioxide, and water. Although the quantities of oxygen and nitrogen in the
atmosphere are very nearly constant, the quantity of
carbon dioxide is slightly variable. Over the land,
the quantity is somewhat less during the day than
during the night. This is doubtless due to the fact
that the assimilation of carbon by plants only takes
place in sunlight. On the other hand, no such
diurnal variation has been observed over the ocean.
The amount of carbon dioxide also varies with the
locality. In the neighbourhood of large towns where
much coal is consumed the amount of carbon dioxide

is slightly greater than in the country. The proportion
of carbon dioxide appears to be nearly independent
of the height above the level of the sea.   Variations
are sometimes observed according to the direction of
the wind.   In crowded rooms the accumulation of
carbon dioxide from respiration may render the air
decidedly injurious, but this vitiated air probably
owes its poisonous properties more to organic im-
purities than to the carbon dioxide itself.

Other Constituents.

The atmosphere contains very variable quantities
of water vapour.   At a given temperature the
atmosphere is incapable of taking up more than
a certain quantity of aqueous vapour, the air is then
said to be saturated with moisture.   The quantity
of water which is capable of being absorbed at a given
temperature is measured simply by the vapour pressure
of water at that temperature.   As is well known, the
vapour pressure of water rises very rapidly with rise
of temperature, and consequently the quantity
water which the atmosphere is capable of absorbing
increases very rapidly with increase of temperature.
The vapour pressure of water has been determined
with great accuracy by Regnault ; a few of his results
are given below.

4—2

Temperature in °C.	Vapour Pressure in mm. mercury.
* − 10°	2·078 mm.
* − 5°	3·113 mm.
0°	4·600 mm.
+ 5°	6·534 mm.
10°	9·165 mm.
15°	12·699 mm.
20°	17·391 mm.
25°	23·550 mm.

* These numbers refer to supercooled water, *not* to ice.

From these numbers we can calculate the quantity of aqueous vapour which a given volume of air at a definite temperature is capable of absorbing. Suppose, for example, we wish to calculate the weight of aqueous vapour which can be taken up by one cubic metre of air at 10° C. At this temperature, the vapour pressure of water is 9·165 mm. One cubic metre of water vapour at normal temperature and pressure weighs 804·75 grams. Hence the weight of aqueous vapour which can saturate this volume of air at 10° C. equals

$$\frac{804 \cdot 75 \times 273 \times 9 \cdot 165}{283 \times 760} = 9 \cdot 362 \text{ grams.}$$

If the temperature of air saturated with moisture falls, the aqueous vapour is precipitated in the form of clouds, rain, snow, or hail according to the atmospheric conditions. Among the conditions which determine the degree of humidity of the atmosphere, distance from large areas of water, and the configuration

of the land are very important. Large quantities
of aqueous vapour are, however, driven by winds into
the interior of continents and more so during summer
than in winter. In discussing the water-content of
the atmosphere, we have two quantities to consider :
firstly, the relative humidity, which is defined as the
ratio of the amount of aqueous vapour present in
a given volume of air at a definite temperature to
the amount which would saturate that volume of air
at the same temperature ; and secondly, the absolute
humidity, which is simply the quantity of water vapour
which is actually present in a given volume of air at
a certain temperature. The amount of aqueous
vapour in the atmosphere diminishes with the height
above the level of the sea, in consequence of the
diminution of atmospheric temperature. It is true,
that inasmuch as aqueous vapour is specifically less
dense than atmospheric air, moist air has a lower
specific gravity than dry air, but the dependence of
the vapour pressure on the temperature is much the
more important factor.

Owing to the putrefaction of organic matter
containing nitrogen, the atmosphere contains minute
quantities of ammonia. But as this gas is extremely
soluble in water, it is very soon removed from the
air by rain water ; in this way it finds its way into the
soil where it soon undergoes nitrification. Traces of
nitrous and nitric acids are also formed by electric

discharges which take place in the atmosphere. Reference has already been made to the occurrence of traces of ozone, and some observers contend that traces of hydrogen peroxide are also present.

The existence of traces of hydrogen in the atmosphere has been proved beyond doubt. Gautier determined the amount of this gas in the atmosphere by passing air previously dried and purified over heated copper oxide and weighing the water formed by the reduction of the oxide to metallic copper. He estimated the quantity of hydrogen at about 0·01 per cent. by volume. On the other hand, Lord Rayleigh has not been able to detect more than 0·003 per cent. of this gas by volume. The hydrogen is probably of volcanic origin but it probably escapes from our atmosphere into space. (See Chapter VII.)

The existence of small quantities of formaldehyde in the atmosphere has been affirmed from time to time by various investigators. In 1904 Henriet determined the amount of this substance in the atmosphere, and obtained the somewhat high result of 2 to 6 grams in 100 cubic metres of air. Gautier pointed out that when air contains considerably less than this quantity of formaldehyde, it ceases to be respirable ; and, in reply, Henriet suggested that the formaldehyde may exist in some form of combination such as methylal.

Atmospheric Organic Matter.

When a strong ray of light passes into a room, a number of minute particles or motes are seen in constant irregular motion. These particles consist partly of inorganic and partly of organic matter. Amongst the organic matter, the germs of plants and animals are always present. The atmospheric microbes are chiefly saprophytic organisms and are the cause of the processes of fermentation and putrefaction, but sometimes pathogenic organisms are present also. Air may be purified from such organisms by passage through asbestos or cotton wool. When purified in this manner, the air no longer exhibits the phenomena of the motile particles when illuminated, the space appearing perfectly empty. Air which has been purified in this manner has been termed optically pure by Tyndall. Organic matter may be preserved for an indefinite length of time in optically pure air without undergoing putrefaction.

When organic matter undergoes putrefaction, various volatile products are liberated, and to such the unhealthiness of closely inhabited spaces is due. The well-known 'stuffiness' of badly ventilated rooms is due not so much to a diminution of the supply of oxygen or to an increase in the quantity of carbon dioxide, but to the poisonous properties of the volatile decomposition products of organic matter which has undergone putrefaction under the influence of micro-organisms. The subject of bacteriology has been enriched by many discoveries of the utmost importance since the pioneer work of Pasteur, but we must refer the reader who desires further information on this subject to some special treatise.

The Rare Gases.

Finally, we must refer to the rare gases present in the atmosphere. In discussing the preparation of nitrogen from the atmosphere, it was stated that under favourable conditions the product obtainable consists of 99 per cent. nitrogen, the one per cent. impurity constituting the inert gases. It is now known that no fewer than five different gases make up this one per cent. impurity. The subject is however of such great interest and importance that a special chapter is devoted to it. (Chapter IX.)

CHAPTER V

MODERN VIEWS ON COMBUSTION

WE have seen that the function of the oxygen of the air is that of a supporter of combustion and respiration. The phenomena of combustion are many and complex, and it is the purpose of the present chapter to give some account of the present state of our knowledge regarding them. This subject has received a considerable amount of attention especially during the last thirty years, particularly as regards the combustion of gases and the aerial oxidation of metals.

Gaseous Combustion.

The gaseous state of aggregation being the simplest one for experimental investigation, it was not unnatural that chemists should devote their attention to gaseous combustion after Lavoisier had released them from the thraldom of phlogistonism. Between the years 1815 and 1825, Sir Humphry Davy carried out a series of experiments on gaseous explosions, primarily with the object of determining the causes of explosions in coal mines, and he discovered the more fundamental facts connected with the ignition of explosive mixtures. He discovered the very important phenomenon of the extinction of flames by cold metallic surfaces, and invented the well-known safety lamp for use in coal mines. In this lamp, the flame is surrounded with fine wire gauze, through which it obtains its supply of oxygen. But the flame cannot readily pass through the gauze, as the high thermal conductivity of the latter cools an external explosive mixture of gases below its ignition-point. Davy also discovered the phenomenon of the flameless combustion of hydrogen and coal gas in contact with a glowing strip of platinum. Little progress was made immediately after Davy's time, until about fifty years later when Bunsen commenced an elaborate investigation of the subject. Although many of Bunsen's results have not been confirmed in their entirety by subsequent investigators, his researches were of great importance, and chemists are indebted to him for having introduced accurate methods of gas analysis. Bunsen was the first to attempt to measure the rates at which flames are propagated in gaseous mixtures; a subject which has been developed enormously in recent years by the work of Berthelot, Mallard and Le Chatelier, Dixon and others.

The earlier investigators recognised a certain temperature which they termed the 'ignition-point' of gaseous mixtures, below which, apparently, it is impossible for the gases to combine, above which combination takes place with explosion. In course of time, however, it was shown that chemical change may be observed to take place in a gaseous mixture at temperatures far below the ignition-point. If the temperature of a mixture of oxygen and hydrogen in a closed vessel be gradually raised, the formation of steam may be observed at temperatures far below that at which explosive combination results. Attempts to determine the ignition-points of various gaseous mixtures have been made in this way; but accurate results cannot be obtained, since the precise temperature at which explosion occurs obviously depends on the amount of slow combustion which has previously taken place. The only way of determining ignition points with accuracy is either to make the preliminary heating-up period almost infinitely short or, better, to heat the combustible gas and air (or oxygen) to the ignition-temperature separately before allowing them to mix. This latter method was employed by Dixon and Coward, in 1909, who obtained the following results at atmospheric pressure :

| | Ignition-temperature °C. | |
Gas	In Air	In Oxygen
Hydrogen	580° to 590°	580° to 590°
Carbon monoxide (moist)	644° to 658°	637° to 658°
Acetylene	406° to 440°	416° to 440°
Methane	650° to 750°	556° to 700°
Ethylene	542° to 547°	500° to 519°
Cyanogen	850° to 862°	803° to 818°

It will be observed that in the case of hydrogen, carbon monoxide, and acetylene, the temperatures of ignition are practically identical in air and in oxygen ; in the case of methane, ethylene, and cyanogen, the ignition-temperature in oxygen is lower than in air.

During 1906 and 1907, Falk, acting on an ingenious suggestion by Nernst, attempted to determine the ignition-points of various mixtures of hydrogen and oxygen, by compressing them in a steel cylinder by means of a weight falling on a piston. In carrying out these determinations, Falk assumed, firstly, that the mixture was heated solely by adiabatic[1] compression, and uniformly throughout its entire mass until the ignition-point was reached, secondly, that the whole mixture then detonated instantaneously, and thirdly, that the piston had no time to descend appreciably after the ignition-point had been reached.

Professor Dixon in his presidential address to the Chemical Society in 1910 criticised Falk's assumptions, and contended that while the first assumption is probably true, the second and third cannot be allowed. In some experiments by Falk's method, Dixon determined the ignition-temperature of a mixture of hydrogen and air (two volumes of the former to five of the latter) and took precautions to arrest the descent of the piston the moment the gases were brought to their true ignition-point. He obtained results in fair agreement with previous determinations at atmospheric pressure. He further showed that when the piston was allowed to descend until it was stopped by the explosion of the gases as in Falk's experiments, the ignition-point was found to be too high, and, further, that if the velocity of the piston

[1] An adiabatic process is one in which no heat enters or leaves the system. It is to be distinguished from an isothermal process which is characterised by the temperature of the system remaining constant.

was increased to a sufficient extent, the apparent ignition-point could be raised as high as 900° C.

The Influence of Moisture on Combustion.

As early as the year 1794, Mrs Fulhame pointed out that the presence of water is essential for certain reactions to take place, in particular for the oxidation of carbon monoxide to the dioxide. This discovery appears to have been forgotten until the phenomena were rediscovered in 1880 by Dixon. Among other things, Dixon discovered that a mixture of carbon monoxide and oxygen when very thoroughly dried does not explode when sparked in the usual way, but on the admission of a trace of water vapour or of some other gas containing hydrogen, combination takes place with explosive violence. This highly important discovery was followed up by some interesting observations by Baker, who found that highly purified sulphur or phosphorus could be distilled in an atmosphere of very dry oxygen, without any combustion occurring. On the admission of a trace of moisture, however, vivid combustion occurred. It was also shown that very carefully purified and dried gaseous mixtures, which in ordinary circumstances readily undergo inflammation, are extraordinarily inert. The amount of moisture necessary to bring about such chemical changes is exceedingly small. It has

been estimated that considerably less than four molecules of steam per one thousand million molecules of
gas are necessary in order to inhibit chemical action.
As regards the explosive combustion of hydrogen and
oxygen, the work of Dixon has shown that the reaction
is probably a direct one, and does not take place
through the intervention of steam for three reasons,
viz., firstly, that well dried mixtures always explode
with a spark, secondly, that the velocity of explosion
in a well dried mixture is greater than when steam is
added, and thirdly, the explosive wave is propagated
as a pressure wave through the mixture. The question
whether water intervenes in the initial interaction of
hydrogen and oxygen below the ignition-point[1] is
a somewhat different one, and can scarcely be regarded
as definitely settled.

Various hypotheses have been framed from time
to time to account for this remarkable catalytic action
of water vapour, but none can be regarded as satisfactory.   We desire, however, to call attention to the
views of Professor Armstrong. Armstrong has always
contended that chemical action cannot take place
between two perfectly pure substances ; a third
substance must be present (even though only in
infinitesimal amount) to form a system which is

[1] Le Chatelier has defined the ignition-point of gases as that
temperature at which the initial flameless combination heats up the
gas, more or less rapidly, until it inflames.

capable of conducting electricity. Armstrong has
termed chemical combination reversed electrolysis.
The function of traces of water vapour is thus readily
intelligible in terms of Armstrong's hypothesis. There
are, however, many grave objections to any such
hypothesis; in particular, there are some well
authenticated cases of chemical combination taking
place between substances which have been most
thoroughly purified and dried. In such cases, Armstrong contends that the fact of chemical combination
having taken place is sufficient to demonstrate the
presence of the necessary impurity. It is perhaps
superfluous to point out to the reader that such
a method of reasoning is irrefutable. A hypothesis
which certainly makes severe demands on the im-
agination was put forward by Baker a few years ago
in which he assumed that water vapour condenses
on certain nuclei, and that the drops thus formed
facilitate chemical change in the layer of gas in
immediate contact with them.

The Combustion of Hydrocarbons.

In former years there was much controversy
regarding the manner in which a hydrocarbon
(such as marsh gas or acetylene) is attacked by
oxygen in combustion. It was widely believed that
the process consisted in what has been termed the
'preferential' combustion of hydrogen: in other

words, the hydrocarbon is dissociated by raising its temperature in air, the hydrogen is burnt first, and the carbon afterwards. A few chemists, on the other hand, considered that the process involved the selective combustion of carbon; others again believed in a simultaneous oxidation of carbon and hydrogen. Doubt, however, arose as to the probability of any idea of preferential oxidation, and the researches of Professor Bone and his collaborators between 1902 and 1906 have completely disproved any such idea. It is impossible in a brief account of the subject such as this chapter attempts to give to do justice to this important work; references to other sources of information will be found in the Bibliography. Experiments were devised for isolating the products of the slow combustion of hydrocarbons and as a result of these experiments it was found that certain substances known as aldehydes were formed during combustion. The formation of aldehydes is exceedingly difficult to reconcile with any other view than that the carbon and hydrogen are attacked simultaneously. These investigators discovered the interesting fact that the combustion of hydrocarbons is not dependent on the presence of traces of moisture; if anything, the velocity of combustion was slightly greater with thoroughly dried gases.

The investigations so far considered were all carried out at temperatures below the ignition-point.

Experiments were next carried out on the explosive combustion of hydrocarbons, and while such phenomena are more difficult to investigate owing to the secondary thermal decompositions which undoubtedly take place at higher temperatures, the results obtained by Bone and his collaborators afford fairly conclusive evidence against the idea of selective combustion. In particular it has been shown that aldehydes are formed in hydrocarbon flames and in explosions.

It is important to observe that the isolation of a product which might conceivably function as an intermediate compound does not necessarily demonstrate that the main reaction takes place in that particular way, since the supposed intermediate compound may be really a by-product of the reaction. Although the isolation of aldehydes in the combustion of hydrocarbons makes it appear highly probable that these substances are the result of the direct interaction of the hydrocarbon and oxygen, there is another possible interpretation of the facts. The hydrocarbon and oxygen may associate with formation of some unknown molecular complex, and this molecular complex decomposes into a number of products including aldehydes. In support of Bone's interpretation of his own results, however, it must be urged that it is probable from kinetic considerations that the primary action between two gases is bimolecular, one molecule of the hydrocarbon reacting with one molecule of oxygen.

Slow Combustion or Autoxidation.

Many interesting phenomena have been observed in the slow oxidation of certain substances, especially metals and organic compounds at ordinary temperatures ; in particular, the phenomena grouped under the term *autoxidation* claim our attention. Between 1858 and 1868 Schönbein made a number of interesting experiments on such processes, and discovered that in a number of cases, one part of the oxygen present combines directly with the substance undergoing oxidation, while another part of the oxygen is converted into ozone or hydrogen peroxide or disappears in some secondary reaction. In some cases he was able to demonstrate the fact that the oxygen apparently distributes itself equally in two ways ; on the one hand, it is consumed by the oxidisable substance, and on the other hand, it becomes 'activated,' being converted into ozone or hydrogen peroxide. Schönbein assumed that this equal division of the oxygen takes place in all cases of slow oxidation ; but owing to secondary reactions, it is only in comparatively few cases that definite quantitative relations can be realised experimentally.

This subject has received attention from a number of investigators, among whom must be mentioned

Moritz Traube, Engler, Manchot, Bodländer, and others. Various hypotheses have been framed from time to time to explain the observed facts, but it would appear that the phenomena have not yet been investigated with sufficient thoroughness for any general theory to be put forward. In particular, it must be observed that there are certain well-known examples of slow oxidation in which 'activation' of oxygen is exceedingly difficult to demonstrate.

The Rusting of Iron.

No account of the chemical relations of the atmosphere would be complete without some description of the present state of our knowledge of a problem of such very great commercial importance as the rusting of iron. The literature on this subject is extensive and controversial, and it would appear that even at the present time, the phenomena are only imperfectly understood. About fifty years ago, it was stated by Crum Brown and also by Crace Calvert that the process is not a simple oxidation of the iron in presence of atmospheric moisture, but is dependent on the presence of carbon dioxide. Rust is an ill-defined mixture of basic carbonates and hydrated oxides of iron, and was supposed to be produced in some such manner as is suggested by the following equations :

(i) $Fe + CO_2 + H_2O = FeCO_3 + H_2,$

(ii) $4FeCO_3 + 6H_2O + O_2 = 4Fe(OH)_3 + 4CO_2.$

and for many years this theory was generally accepted.

However, in 1905, the subject was reinvestigated by Dunstan, Jowett, and Goulding, and these investigators concluded that the process of rusting involves a direct interaction between iron, water, and oxygen. They expressed the view that the part played by atmospheric carbon dioxide is quite unimportant, and that the formation of hydrogen peroxide is an essential part of the process. Although their experiments failed to demonstrate the presence of hydrogen peroxide, they regarded its formation as an intermediate product a very essential part of the process, basing their argument chiefly, if not wholly, on the fact that the addition of substances which decompose hydrogen peroxide inhibits the process of rusting. It was also shown that when iron was immersed in a solution of hydrogen peroxide, rapid oxidation ensued with the formation of a substance of the approximate composition $Fe_2O_2(OH)_2$.

In the following year, Moody made an elaborate investigation of the rusting process. He criticised the work of Dunstan and his collaborators very adversely, and pointed out quite correctly that in their most important experiment, adequate precautions to ensure the absence of carbon dioxide from the apparatus were not taken. Dunstan and his co-workers had observed the production of a green colour prior to the actual appearance of rusting, and Moody remarked that 'no more conclusive evidence that carbonic acid was present in the materials used could be afforded than the production of this green colour which invariably accompanies the early attack of carbonic acid on iron in presence of air or oxygen.'

Moody devised an apparatus capable of leaving iron, oxygen, and water in contact for many weeks without a speck of rust appearing on the surface of the metal. When air is drawn through this apparatus (Fig. 2) by means of the aspirator G, it is thoroughly freed from carbon dioxide before reaching the iron in the bend C. From and including the tap N to the side of the tube joining E and F all intermediate parts of the apparatus are united by sealed glass junctions, so that with ordinary precautions, access of atmospheric carbon dioxide to the iron in C is entirely prevented.

The iron used in these experiments was a soft Swedish iron containing 99·8 per cent. of iron, and each piece used weighed about 0·9 gram. In the method of procedure finally adopted, it was found necessary to prevent contact between the iron with the glass of the tube by partly coating it with paraffin wax, and also to cover the iron with a one per cent. solution of chromium trioxide before sealing the tube into position. Air was then drawn through the apparatus for three weeks, and the chromic acid

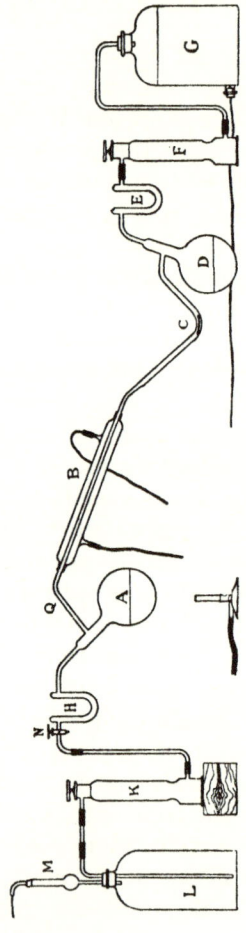

Fig. 2. Moody's apparatus. A distilling flask, B condenser, C iron in bent tube, D receiving flask, E soda lime tube, F caustic potash tower, G aspirator, H soda lime tube, K caustic potash tower, L air reservoir, M soda lime tube.

removed from the iron by distilling a large quantity of
water from the flask A (Fig. 2) which contained a one
per cent. solution of barium hydroxide. When the iron
had been washed quite free of chromic acid, air was
again drawn through the apparatus for about three
weeks. At the end of this time the iron remained
perfectly bright. But on drawing moist air containing
carbon dioxide through the apparatus, rusting appeared in six hours. It may therefore be concluded
that when carbon dioxide is rigidly excluded, no
interaction takes place between iron and oxygen in
presence of water.

In the course of his investigation, Moody examined
a number of specimens of rust and found that every
one of them contained ferrous carbonate in varying
proportions. This fact renders it extremely improbable that hydrogen peroxide is an intermediate
product in the rusting process, since both ferrous
oxide and carbonate are at once oxidised by this
substance. The probable interpretation of the phenomena is that the first stage of atmospheric corrosion
involves the formation of ferrous carbonate which
gradually becomes basic owing to loss of carbon
dioxide.

With regard to Dunstan's experiments on the
oxidation of iron by hydrogen peroxide, Moody has
shown that when the peroxide is carefully purified,
no rusting occurs although oxygen is freely evolved

at the surface of the metal. In 1907 Dunstan repeated Moody's experiments and found that rusting did take place even although carbon dioxide was apparently excluded.

In 1908 Sir William Tilden repeated Moody's original experiment with a slight modification. Instead of placing one piece of iron in the bend of the tube, three pieces were placed in separate bulbs blown on one piece of tube. The first was a piece of iron partly polished and partly rough filed. The second had been previously heated in hydrogen to remove the film of oxide, and subsequently to redness *in vacuo*. The third was a piece of the same iron which had been immersed in a one per cent. solution of chromium trioxide for twenty-four hours and afterwards carefully washed. Each piece of iron was protected from contact with the glass by the aid of paraffin wax as in Moody's experiments. Tilden found that while the pieces of iron which had not been treated with chromic acid rapidly became coated with rust, the piece which had been treated in this way remained bright for a long period. When iron has been immersed in certain reagents it becomes immune or 'passive' to many solvents. It would appear therefore that the previous treatment of the iron with chromic acid introduces complications into the problem by passifying the iron. Tilden concluded that commercial iron, liquid water, and oxygen are alone sufficient for the production of rust : carbon dioxide is not necessary, but when present hastens the action.

In 1911 Dunstan and Hill returned to the subject and explained Moody's result as due to the passifying of the iron by the chromic acid and its subsequent activation by carbon dioxide. They also described some experiments which have apparently demonstrated the fact that hydrogen peroxide is actually produced during the process of rusting.

We may therefore sum up the present state of our knowledge of the subject by stating that as far as the rusting of commercial iron is concerned, the question of the necessity of the presence of carbon dioxide is an entirely open one ; and it appears to the present writer that the only way of deciding the matter would be to repeat Moody's experiment, but instead of using chromic acid to employ some inert organic liquid, which exerts no solvent action on the paraffin wax, with which to cover the iron during the process of sealing the tube into position.

So far the experimental work which we have been considering has been carried out on iron of a moderate degree of purity, i.e. on a material consisting of over 99 per cent. of pure iron.   From the chemical point of view, however, the amount of impurity in such iron is far from negligible, and in 1910 a research was carried out by Lambert and Thomson in which extraordinary precautions were taken to prepare iron of a very high degree of purity, and to determine the conditions which are necessary for such iron to rust. Vessels of silica which had been very thoroughly cleansed were employed to contain the iron during the experiments. These investigators found that their pure iron did not undergo any visible oxidation when treated with pure. water and pure oxygen in vessels of clear fused silica. But when iron of a lesser degree of purity was employed, oxidation took place in two or three hours. It is perhaps necessary to remind the reader that while the work of Lambert and Thomson is of very great interest, it cannot claim to have the same technical value as the earlier investigations of Moody and others.

Professor Armstrong believes that the rusting of iron is essentially an electrolytic phenomenon. Since pure water is not an electrolyte, he inclines to the view that Moody's theory is essentially correct, the carbon dioxide dissolving in the water and imparting to it the necessary electrolytic property. Further, for a metal to go into solution in an electrolyte, it is necessary for some less electro-positive metal to be present. In the process of the rusting of ordinary iron, the small quantity of impurity is more than sufficient to furnish this less electro-positive metal. The non-rusting of Lambert and Thomson's highly purified iron would therefore be easy to understand. But it must be borne in mind that silica vessels were employed to contain the iron. So far as the present writer is aware, these investigators did not make any special experiments to ascertain whether highly purified water has any solvent action on silica[1]. If it should be found that such is the case, we have a complete explanation of their results; the water with traces of silicic acid in solution constitutes the electrolyte, no electro-negative element was present in the experiments with the highly purified iron, and therefore rusting did not take place. But with the less pure iron, all the necessary conditions according to Armstrong's theory are fulfilled, and the formation of rust was observed.

CHAPTER VI

CONSTANCY OF THE COMPOSITION OF
THE ATMOSPHERE

In Chapter IV we dealt with the constituents of the atmosphere. It is the purpose of the present chapter to give some account of the methods of determining the more important of these constituents

---

[1] Since the above was written, a paper has appeared by Lambert (*Proc. Chem. Soc.* 1912, p. 197) in which the author states *inter alia* the special experiments have demonstrated the fact that quartz does not dissolve to a sufficient extent in water to invalidate the conclusion previously reached.

quantitatively, and of some of the results which have been obtained. Generally speaking, the amounts of nitrogen and oxygen are very nearly constant; the carbon dioxide undergoes slight variations, while some of the other constituents, especially the aqueous vapour and organic matter, vary very considerably in amount.

At first sight it may strike the reader that the constancy of the composition of the atmosphere is a matter of comparatively little interest. But it must be borne in mind that the atmosphere is not inert, for chemical changes on an enormous scale are constantly taking place, and the constituents of the atmosphere differ greatly in density. It is true that a mixture of gases of very different densities left to itself in a closed vessel shows no tendency towards segregation: the process of diffusion, due to the movements of the individual molecules, which is constantly taking place, results in the composition of the mixture being identical at every point. One may be tempted to apply these results of a small scale experiment to the atmosphere, but diffusion is such a slow process that the atmosphere possessing a nearly constant composition is all the more remarkable, even though gaseous interchange is aided by winds.

The first accurate determinations of the ratio of oxygen to nitrogen in the atmosphere were made by Cavendish in 1781. He determined the volume composition of the air by adding known volumes of hydrogen, causing the oxygen and hydrogen to combine under the influence of the electric spark, and observing the resulting diminution of volume. He collected samples of air from different localities and at different times of the year in order to ascertain whether it was 'more phlogisticated' at one time than another. Cavendish concluded that the composition of the atmosphere as far as the amount of oxygen was concerned was invariable, and the result which he obtained, 20·83 per cent. of oxygen by volume, is remarkably close to that obtained by the best determinations of modern observers.

The fact that the atmosphere consists approximately of four parts of nitrogen to one part of oxygen by volume led several chemists to maintain that air is a definite compound of these two gases in the above ratio. That such an idea is erroneous follows from a number of facts which it may not be out of place to summarise here. In the first place, the properties of the atmosphere are the average of its constituents; in general, the properties of a compound are markedly different from those of its constituents. Secondly, inasmuch as oxygen is more soluble in water than nitrogen, when water is agitated in contact with air the oxygen is absorbed to a greater extent than the nitrogen. Thirdly, when nitrogen and oxygen are mixed in the proportions in which they exist in the atmosphere, no thermal change is observed. Lastly, when air is liquefied, the nitrogen, being more volatile than the oxygen, boils off first, and a separation of the constituents may be thereby effected.

The ratio of oxygen to nitrogen in the atmosphere has formed a subject for investigation by many chemists. The older determinations by Gay-Lussac, Davy, and others, yielded numbers of the order of 21 per cent. by volume of oxygen but are not very accurate. The first really accurate volumetric determinations were made by Bunsen in 1846. Bunsen employed Cavendish's original method of sparking with hydrogen, but he greatly improved the apparatus for this purpose. In Bunsen's method, the gases are confined over mercury, and the measurements of temperature, pressure, and volume are effected with great accuracy. Bunsen found that the mean percentage of oxygen is 20·924 parts by volume.

The work of Regnault, Morley, Leduc, and others has shown that the oxygen-content of the atmosphere undergoes slight but quite perceptible variations. The extreme limits are from 20·4 to 21 per cent.

A very extensive series of analyses of air from various sources has been carried out by Angus Smith. The following results are taken from his work *Air and Rain* :

Oxygen in the air		Volume per cent.
North-east sea shore and open heath (Scotland)	...	20·999
In the outer circle of Manchester, not raining	...	20·947
Low parts of Perth	...	20·935
London, open places, summer	...	20·950
Pit of theatre. 11.30 p.m.	...	20·740
Court of Queen's Bench, February 2nd, 1866 ...	...	20·650
Under shafts in metalliferous mines	...	20·424

The general conclusion arrived at by Angus Smith is that the air of mountains and moorlands is richer in oxygen than that of towns.

The late Sir Edward Frankland made some analyses of air taken from various parts of Mont Blanc in an ascent made in 1859. The following results were obtained for the percentage of oxygen :

At Chamonix (altitude 3000 feet)	20·881 per cent. by volume
At Grands Mulets (altitude 11,000 feet)	20·779 ,, ,,
At the Summit (altitude 15,732 feet)	20·950 ,, ,,

'So far as the nitrogen and oxygen are concerned, the composition of these samples of air falls within the limits of variation noticed by former experimenters....' (Frankland, *Researches*, p. 476.)

The results which we have been considering so far have been obtained by the volumetric method of explosion with hydrogen, but interesting results were obtained by a gravimetric method by Dumas and Boussingault in 1841. The principle of the method employed by these chemists was to pass air, freed from carbon dioxide and aqueous vapour, over

metallic copper heated to redness. In this way
all the oxygen was removed as copper oxide, the
nitrogen being collected in a glass bulb. From the
increase in weight of the tube filled with copper, and
of the bulb, before and after the experiment, the
percentage of oxygen and nitrogen could be directly

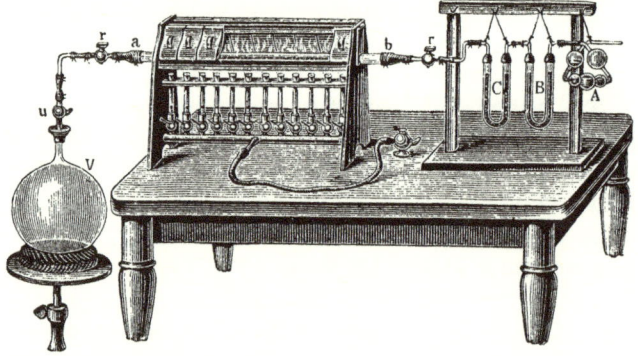

Fig. 3. Dumas and Boussingault's apparatus. *A* potash bulbs, *B* and
C tubes containing pumice moistened with strong sulphuric acid,
ab tube containing copper heated to redness in the furnace,
V exhausted bulb for collecting the nitrogen.

determined. The mean results obtained by the gravimetric method were :

Oxygen	...	23·0 per cent. by weight.
Nitrogen	...	77·0 ,, ,,

These results are in agreement with those obtained
by the gasometric method, when the numbers are

reduced to volume composition from the known densities of the two gases. The gravimetric method although capable of giving accurate results is exceedingly tedious and requires complicated apparatus. The volumetric method is very much more convenient and rapid and is capable of yielding results of great accuracy. From what has been stated above, it is clear that the nitrogen (with the inert gases) is always determined by difference.

We must now consider the methods for determining the percentage of carbon dioxide in the atmosphere. The first method is a gravimetric one. It depends upon the fact that carbon dioxide is absorbed by caustic potash with formation of potassium hydrogen carbonate. The method consists in drawing a known volume of air, carefully dried by a substance such as concentrated sulphuric acid, through tubes containing caustic potash, the carbon dioxide being determined by direct weighing. Although apparently so simple, the method requires great care, and a large volume of air must be drawn through the apparatus in order to obtain sufficient carbon dioxide for an exact weighing. A much more rapid and accurate method of determining atmospheric carbon dioxide, and which has the advantage of only requiring a comparatively small volume of air, was devised by Pettenkofer in 1858. This method depends upon the fact that a solution of barium hydroxide when shaken up with air in a closed vessel, abstracts the whole of the carbon dioxide from that volume of air with formation of insoluble barium carbonate. The quantity of baryta remaining in excess is then determined volumetrically by titration with standard oxalic or sulphuric acid.

It has already been stated (Chapter IV) that the amount of carbon dioxide is slightly variable from time to time, and in different localities. The older determinations gave 0·04 per cent. by volume as an average but there can be little doubt that 0·03 per cent. is a more accurate figure. The researches of Angus Smith appear to indicate that the diminutions in the percentage of oxygen in the atmosphere are accompanied by concomitant increases in the percentage of carbon dioxide. Similar results have been obtained by Frankland. In large towns where much coal is consumed and in crowded areas, the percentage of carbon dioxide may rise very considerably. The following results are taken from Angus Smith's *Air and Rain.*

Carbon dioxide in the air	Volume per cent.
On hills in Scotland (1000 to 4406 feet)	0·0332
In London parks and open places	0·0301
Manchester streets in ordinary weather	0·0403
About middens	0·0774
In theatres, worst parts, as much as	0·3200
In mines : largest amount found in Cornwall ...	2·500

Apart from such local variations, and the slight diurnal variation due to the cessation of the assimilatory activity of plants in the night, the amount of carbon dioxide in the atmosphere remains between fairly narrow limits.

The variations in the amount of water vapour in the atmosphere are of very great importance from the meteorological standpoint. Various methods of determining the hygrometric state of the atmosphere have been devised from time to time. The moisture may be determined chemically by absorption in some desiccating material such as concentrated sulphuric acid or phosphoric anhydride. This method is however little used. One form of hygrometer which is frequently employed where no great accuracy is required consists of two similar thermometers side by side, the bulb of one being kept moist by being surrounded with a wet piece of cloth the other end of which dips into a vessel of water. When the air contains but little moisture, evaporation from the wet cloth takes place rapidly, and the temperature of the wet bulb of the thermometer falls owing to the withdrawal of the heat required to vaporise the water. When the atmosphere is thoroughly saturated with aqueous vapour the two thermometers record the same temperature. Various formulae have been devised for connecting the hygrometric state of the air with the difference of temperature as indicated by the two thermometers, but none are very satisfactory. In general the apparatus is used as an indicator rather than as a

measuring instrument. Many other forms of hygrometer have been employed, but we must refer the
reader to some treatise on meteorology for a description of them.

The Composition of the Upper Atmosphere.

We have seen that the composition of the atmosphere up to elevations of several thousand feet is
characterised by great uniformity. It is true that
slight variations in composition have been observed,
but the approximate constancy of composition is very
remarkable. On the other hand, evidence has been
adduced to show that the composition of the atmosphere at very great elevations is markedly different
from that of the lower atmosphere ; in particular, it
is almost certain that the proportions of hydrogen
and helium (Chapter IX) increase considerably at
very great altitudes. In 1912 Wegener, in a most
interesting investigation of the upper layers of the
atmosphere, brought forward strong evidence in favour
of the existence of an upper atmosphere of hydrogen.
The evidence was derived chiefly from calculations of
partial pressure, and supported by observations of
the total reflection of sound waves, of the appearance
and disappearance of twilight, of luminous night clouds,
and of the glowing of meteors. These various lines
of enquiry indicate 70–80 kilometres as the approximate height of the base of the hydrogen stratum. The

outermost layers of the atmosphere according to
Wegener contain an unknown gas lighter than hydrogen, which he termed geocoronium. The following
table gives the relative distribution of the more
important constituents of the atmosphere up to elevations of 500 kilometres. The numbers are expressed
in percentages by volume.

Height (km.)	Atmospheric pressure (mm.)	Geoco-ronium	Hydrogen	Helium	Nitrogen	Oxygen	Argon
0	760	0·00058	0·0033	0·0005	78·1	20·9	0·937
20	41·7	0	0	0	85	15	0
40	1·92	0	1	0	88	10	—
60	0·106	5	12	1	77	6	—
80	0·0192	19	55	4	21	1.	—
100	0·0128	29	67	4	1	0	—
120	0·0106	32	65	3	0	—	—
140	0·00900	36	62	2	—	—	—
200	0·00581	50	50	1	—	—	—
300	0·00329	71	29	—	—	—	—
400	0·00220	85	15	—	—	—	—
500	0·00162	93	7	—	—	—	—

A few words must be added regarding the hypo-
thetical gas geocoronium. Some years ago, Mendeleeff
predicted the existence of two elements both lighter
than hydrogen belonging to the group of inert elements
(Chapter IX). These elements have been provision-
ally termed x and y. Many years previously an
unknown line in the solar spectrum was observed in
the sun's corona during a solar eclipse, which was
attributed to an unknown element termed coronium;
and Mendeleeff sought to identify coronium with the

element y. Mendeleeff predicted an atomic weight of 0·4 for this gas. The spectrum of the aurora contains a well-defined line, not identical with the line attributed to coronium, which according to Wegener is due to the presence of geocoronium. In spite of the difference in their spectra Wegener believes that geocoronium is identical with coronium. The presence of coronium in the lower atmosphere has not been definitely demonstrated, although in 1893 Nasini, Anderlini, and Salvadori considered that they had observed the spectrum of this element in certain volcanic gases. The late Dr Johnstone Stoney pointed out that certain light gases, notably hydrogen and helium, probably escape from our atmosphere into space, and if this is true of these gases it is doubtless true of coronium also. It is therefore perhaps not surprising that coronium has not been discovered in the lower atmosphere.

CHAPTER VII

THE ESCAPE OF GASES FROM PLANETARY ATMOSPHERES ACCORDING TO THE KINETIC THEORY

THE discovery of the quantitative relation between the pressure and volume of a given quantity of gas by Boyle and later by Mariotte in the seventeenth century was followed by the theoretical views of

Daniel Bernoulli in 1738 regarding the behaviour of gases which are essentially similar to those generally held at the present time. Apparently no fresh contribution to the theory of gases was made until the year 1845 when Waterston communicated a paper to the Royal Society containing a very complete development of the kinetic theory. The paper, however, was not published at the time, but on account of its historical interest its publication was secured by Lord Rayleigh in 1892. In 1856 Krönig and in 1857 Clausius independently developed the same theoretical ideas. In more recent times the kinetic theory of matter was greatly elaborated and extended by the work of Clerk Maxwell, Boltzmann, and others.

According to this theory, the ultimate particles of a gas, which are identical with the molecules, are more or less independent of one another, and are in constant rapid motion. The pressure exerted by gases is due to the bombardment of the walls of the containing vessel by the molecules. On collision, the molecules are supposed to behave as perfectly elastic spheres. The path traversed by any individual molecule is an irregular one due to the successive encounters with other molecules. The distance traversed by any molecule between one collision and the next will change at each collision. The average distance traversed by a molecule without collision is termed its mean free path.

If we consider the molecular weight of any gas expressed in grams, it can be shown that

$$pv = \frac{1}{3} Mu^2,$$

where p and v denote respectively the pressure and volume of the gas, M denotes the molecular weight in grams and u denotes the speed of the molecules.

From this equation we can calculate the mean velocity of any gas molecule under known conditions. For example, let us calculate this velocity for the hydrogen molecule at 0° C. and 760 mm. pressure. The temperature 0° C. is equal to 273° on the absolute scale. The mass of 1 c.cm. of hydrogen under these conditions is 0·00009 gram and the absolute value of a pressure of 760 mm. of mercury is $76 \times 13\cdot6 \times 981$ dynes per sq. cm.

Hence

$$u = \sqrt{\frac{3 \times 76 \times 13\cdot6 \times 981}{0\cdot00009}} = 184{,}000 \text{ cm. per second.}$$

The velocity of the molecule of any other gas can be calculated, since this quantity is inversely proportional to the square-root of the molecular weight.

In Chapter VI we referred briefly to the views of Johnstone Stoney regarding the escape of gases from planetary atmospheres. It is now necessary to

examine this subject a little more closely as regards
the kinetic considerations on which Stoney's views
are based and some of the directions in which
this hypothesis requires to be more fully worked
out.

In 1868 Stoney investigated theoretically the
conditions which limit the extent of the atmospheres
of planets and satellites and pointed out that other
things being equal, the less the gravitational attraction
of a planet, the greater will be the height to which
its atmosphere will rise, and further, if the force of
gravity be sufficiently low and the velocity of the gas
molecules be sufficiently high, some of the molecules
will escape from the atmosphere of the planet and
become independent wanderers throughout space.

In 1898 Stoney developed his fundamental views
in somewhat greater detail.   He pointed out that the
boundary between those gases which can effectually
escape from the Earth and those which cannot, lies
between gas with molecules which are twice as heavy
as hydrogen and gas with molecules nine times as
heavy as the molecule of hydrogen.   He regarded the
temperature of the upper limit of the Earth's atmosphere as $-66°$ C. or $207°$ absolute.   We have already
calculated the velocity of mean square of the hydrogen
molecule and found it to be 1840 metres per second
at $0°$ C.   Since the velocity is proportional to the
square root of the absolute temperature, the velocity

of the hydrogen molecule at $-66°$ C. is equal to

$$1840 \sqrt{\frac{207}{273}} \text{ metres per second}$$
$$= 1600 \text{ metres per second.}$$

In a similar way the velocity of the helium molecule at the same temperature can be calculated to be 1133 metres per second. These velocities are average velocities, but the velocity of some molecules greatly exceeds the mean value; and, according to Stoney, a velocity between nine and ten times the mean velocity for these gases is sufficiently great for the molecules of these gases to escape from the Earth's atmosphere, and further, such velocities are sufficiently frequently attained to make the escape of gas effectual.

On the other hand, the velocity of mean square of a molecule of water vapour would require to be accelerated about twenty times before it would exceed the critical velocity. But, according to Johnstone Stoney, the velocity of very few molecules of water vapour would reach the critical velocity and consequently no appreciable quantity of water vapour would escape. This is in accordance with experience; water does not tend to leave the Earth.

By calculating the necessary data for the other planets, Stoney arrived at the following conclusions regarding their atmospheres:

The Moon appears to be devoid of any atmosphere.

Mercury cannot retain any water vapour, and gradually loses oxygen and nitrogen; the only constituents which are likely to remain on this planet are argon (Chapter IX) and carbon dioxide. The atmosphere of Venus is very similar to that of the Earth. As regards Mars, water and the lighter gases cannot remain on that planet, the atmosphere of which probably consists in the main of nitrogen, argon, and carbon dioxide. Since water cannot remain on Mars, the planet is probably devoid of vegetation, and it is therefore improbable that much free oxygen is present. The polar snows of Mars probably consist of solid carbon dioxide. Jupiter is able to retain all gases known to chemists.

As regards the atmospheres of Saturn, Uranus, and Neptune, Stoney expressed the view that as far as the available data were able to lead, the atmosphere of Saturn is similar to that of Jupiter, while the atmospheres of Uranus and Neptune resemble that of the Earth.

In 1899 S. R. Cook criticised Stoney's discussion of the problem on the ground that Stoney had omitted to determine by the aid of the kinetic theory the relative number of molecules which would have a velocity sufficient to enable them to escape from the atmosphere of the Earth or other planet, and by applying Maxwell's law of the distribution of velocities to the subject he concluded that the Earth and the major planets would be able to retain hydrogen and helium.

In reply to this criticism, Stoney contended that the data furnished by the kinetic theory are inadequate, and pointed out that the Moon has not retained an atmosphere and that the Earth and Venus do retain aqueous vapour in their atmospheres. Stoney pointed out that Maxwell's law fails just where its aid is required, namely in the outer region of an atmosphere where the escape of molecules is taking place.

In 1900 Professor G. H. Bryan criticised Stoney's reasoning on similar grounds and Stoney replied in much the same terms as before, pointing out that Maxwell's law is inapplicable to the outermost strata of an atmosphere ; and further that in such strata which are characterised by an almost total absence of molecular encounters, any law governing the distribution of velocities would be utterly unlike Maxwell's law or any of its modifications.

In 1901 Rogovsky, in a theoretical investigation of the temperatures and compositions of planetary atmospheres, pointed out that the temperature $-66°$ C. assumed by Stoney as the temperature of the upper limits of an atmosphere is incorrect. He pointed out that a temperature of $-75°$ C. has been observed at an elevation of 11,000 metres above the surface of the Earth, and concluded that the maximum densities of

the gases which are not retained in the upper strata
of the atmospheres of planets as determined by Stoney
are too high.   In particular, he expressed the view
that helium and possibly hydrogen may be retained
in the upper strata of the Earth's atmosphere.

It is clear, therefore, that Stoney's hypothesis
requires to be worked out much more completely
than it is at present, although it will probably be
generally admitted that the hypothesis is correct as
far as the fundamental ideas are concerned.   With
improved methods of spectroscopic investigation,
we may hope that our knowledge of planetary atmospheres will be greatly increased.

CHAPTER VIII

LIQUID AIR

THE earliest experiments on the liquefaction of
gases were made by Davy and by Faraday about the
year 1823.   The method adopted was to employ
∧ shaped sealed glass tubes in one limb of which
were placed the substances from which the gas could
be obtained by heating the tube.   The other limb
was cooled in a freezing mixture.   Under the influence
of the cold and pressure, a number of gases were

condensed to the liquid state. But several gases, notably hydrogen, oxygen, nitrogen, and carbon monoxide, resisted all attempts at liquefaction, and such gases were termed permanent gases by Faraday.

In 1834 Thilorier liquefied carbon dioxide by cold and pressure, and by allowing the liquid to evaporate rapidly he succeeded in solidifying the substance. The solidification is due to the withdrawal of heat from the liquid by its own rapid evaporation.

Faraday's idea of permanent gases was universally accepted until the year 1869, when Andrews carried out some elaborate experiments on the compressibility of carbon dioxide. It had been previously shown in 1822 by Cagniard de la Tour that when a liquid is heated in a closed vessel, it vaporises completely above a certain temperature, whatever the amount of liquid. In this work we can see the germ of the discovery of a critical temperature. Faraday appears to have regarded the so-called permanent gases as being in a state comparable to liquids which are completely vaporised as in the Cagniard de la Tour experiment, but he does not appear to have attempted to push the enquiry far enough to verify his views by experiments on the gases which he actually liquefied. The work of Andrews showed that there is a certain temperature, $30\cdot92°$ C. in the case of carbon dioxide, above which it is impossible to liquefy the gas by the application of any pressure however great. Andrews

further showed that it is possible to transform
carbon dioxide from the liquid to the gaseous state
and *vice versâ* without any breach of continuity.

In 1877 Pictet and Cailletet, independently, and
by totally different methods, succeeded in liquefying
oxygen. Pictet employed the method of continuous
cooling. Carbon dioxide was cooled and condensed
by means of liquefied sulphur dioxide, and this liquid
carbon dioxide was circulated round a long cylinder
in connexion with an apparatus for generating oxygen.
The oxygen was cooled below its critical temperature
by the evaporation of the liquefied carbon dioxide,
and was suddenly allowed to expand after having
been subjected to a pressure of about 320 atmospheres.
When a gas is suddenly allowed to expand, it does
work against the external pressure, and this work is
done at the expense of its own heat with the result
that its temperature falls. In this way Pictet succeeded in cooling the oxygen to its liquefaction point.

Cailletet employed a much simpler form of apparatus. He confined the oxygen in strong glass
tubes over mercury, and after applying a pressure of
400 atmospheres, he suddenly released the pressure.
The fall of temperature was so great that the critical
temperature was passed and the oxygen appeared as
a spray in the tube. In Cailletet's method the cooling
of the gas was effected by the performance of external
work alone.

After the researches of Pictet and Cailletet, the liquefaction of gases received attention from a number of investigators, among whom mention must be made of Wroblewski, Olszewski, Kamerlingh Onnes, and Sir James Dewar. The last named has made a number of important investigations on the physical and chemical properties of matter at low temperatures.

It has been shown thermodynamically that if a perfect gas, i.e. one which obeys the gas laws exactly, is allowed to expand without doing external work, no reduction of temperature would occur. No gas however obeys those laws exactly, and an important investigation carried out many years ago by Joule and Lord Kelvin showed that when a steady stream of gas at a constant temperature was allowed to flow through a porous plug in such a manner that its velocity before and after passage through the plug is small, the temperature of the issuing gas was slightly lowered. In all the gases which they investigated, this slight reduction of temperature was observed, except in the case of hydrogen, in which a slight increase of temperature was observed after passage through the plug. This phenomenon, known as the Joule-Thomson effect, has been employed successfully in the liquefaction of air on the commercial scale. In the experiments of Joule and Kelvin, the fall of pressure was not great, and the change of temperature was only a fraction of a degree Centigrade, but by

employing greater differences of pressure the magni-
tude of the Joule-Thomson effect is correspondingly
increased.

In 1895 Linde in Germany and Hampson in this
country applied the Joule-Thomson effect to the
construction of apparatus for the production of liquid
air on a technical scale. The principles of both are
identical, the only differences are in the details.   In
Hampson's apparatus, the air, freed from moisture
and carbon dioxide, is compressed to about 180 atmospheres and then allowed to escape through a small
orifice. In this way the temperature of the issuing
air is lowered. The air which escapes is made to
sweep over the tube which delivers it. In this way
the cooling effect becomes cumulative until finally
the air liquefies and can be drawn off. The apparatus
is an extremely convenient one and the change of
temperature observed is due to the performance of
internal work. The yield of liquid air is 1 to $1\frac{1}{2}$ litres
per hour.

Some years elapsed before the liquefaction of
hydrogen was successfully accomplished. Dewar first
liquefied this gas in 1898 by compressing it to 200
atmospheres and simultaneously cooling it to a temperature below $-200°$ C.   In those circumstances the
sign of the Joule-Thomson effect changes, and hydrogen behaves like other gases in passing through a fine
orifice, becoming cooled. Dewar has succeeded in

obtaining liquid hydrogen in bulk. In 1901 a simple method of obtaining liquid hydrogen in quantity, depending on the same principles, was devised by Travers.

For some years helium remained the only gas which resisted liquefaction, but in 1908 Kamerlingh Onnes effected the liquefaction of this gas also. In this research, the nearest approach to the absolute zero of temperature[1] was reached. Onnes found that the boiling point of the liquid was $4 \cdot 3°$ absolute. Evaporation under diminished pressure did not lead to solidification, although a temperature of $3°$ absolute was reached. In 1911 Onnes succeeded in realising the very low temperature of $1 \cdot 48°$ absolute.

With the exception of helium, every other gas has been solidified by allowing the liquid to evaporate rapidly. It is remarkable that oxygen, which is more easily liquefied than hydrogen or nitrogen, was not solidified till 1911, by rapid evaporation of the liquid under greatly reduced pressure.

The manipulation of liquefied gases has been rendered a simple matter by Dewar's invention of the well-known vacuum vessels. The space between the walls must be exhausted to a very high vacuum, since the conduction of heat by gases over

---

[1] The absolute zero of temperature, $-273°$ C., must be regarded as a definite temperature, as there are a number of lines of enquiry which lead to this temperature.

a considerable range of pressure is independent of the pressure, but at very low pressures is proportional to the pressure. In such vessels, the chief source of the transference of heat to the interior is radiation, which can be reduced to a minimum by silvering the walls. In a good vessel, it is possible to preserve liquid air for several days.

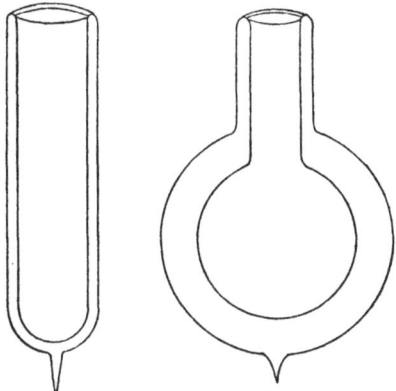

Fig. 4. Vacuum Vessels.

When air is freshly liquefied by Hampson's apparatus, it consists of about equal parts of oxygen and nitrogen. But as nitrogen at normal atmospheric pressure boils at about −195° C., and oxygen at −182° C., the nitrogen boils away more rapidly than the oxygen, and the final dregs of liquid consist almost entirely of oxygen with traces of nitrogen, argon and its heavier congeners. Since the temperature of liquid air rises rapidly during its evaporation until the boiling point of oxygen is reached, it is important to have a reliable and accurate method of determining the temperature at any stage. The temperature can always be determined by the aid of a platinum resistance thermometer, or by a thermo-couple, or by a hydrogen thermometer, or by making an analysis of the vapour which is being given off. This last method, which was worked out by Mr Baly, has been carefully checked by the hydrogen thermometer and should prove of great value for checking resistance thermometers for low temperature research. For the measurement of still lower temperatures satisfactory results have been obtained by the use of the helium thermometer.

Liquid air is a clear mobile liquid with a faint blue colour. The colour is due to the oxygen, and increases in intensity as the nitrogen boils off. When poured into water, some drops sink to the bottom while others rise to the surface, and become surrounded by envelopes of gaseous air and thin elastic scales of ice. Many substances when subjected to the low temperature of liquid air undergo extraordinary changes of properties. Indiarubber for example becomes hard and exceedingly brittle. A finger may be thrust into liquid air for a moment without danger, as a thin film of gaseous air prevents actual contact; but if left in the liquid for more than a moment, severe pain is caused by the low temperature.

The production of liquid air in bulk has been of the greatest importance in the advancement of our knowledge of the properties of matter at low temperatures. In 1892 Dewar and Fleming showed that the electrical resistance of pure metals decreased steadily with diminution of temperature, and in such a manner as to suggest that if the absolute zero of temperature could be reached the resistance would disappear altogether. Further experiments at the very low temperatures obtained by liquid hydrogen appeared to indicate that at absolute zero the electrical resistance would not vanish but simply approach a limiting value, but the recent work of Kamerlingh Onnes at extremely low temperatures does indicate the disappearance of resistance at absolute zero. Many substances display interesting optical phenomena at low temperatures; in particular, many substances which are not fluorescent or phosphorescent at ordinary temperatures, show these effects strongly at very low temperatures. In 1904 Dewar showed that charcoal at the temperature of liquid air becomes a very efficient absorbent of gases, and this discovery has been of great value in the study of the inert gases. Liquid air has also been of great value in physiological research; the study of the action of low temperatures on various organisms has already yielded results of great interest. But perhaps the most interesting phenomena observed at low temperatures are those relating to the cessation or suppression of chemical action. Most chemical reactions possess an exceedingly high temperature coefficient of velocity, and reactions which proceed vigorously at ordinary temperatures are almost suppressed at very low temperatures. There is, however, one remarkable exception to this: fluorine retains its activity at exceedingly low temperatures. Moissan and Dewar found that fluorine and hydrogen unite with explosive violence at a temperature of $-252°$ C.

Finally we may refer to the use of liquid air for drying gases. The usual method of obtaining a well-dried gas is to pass the gas over some desiccating material such as phosphoric anhydride, but for some purposes the use of this substance is inadvisable. In such circumstances the gas may be efficiently dried by passing it through a U tube cooled in liquid air. The water vapour is thereby frozen out, its vapour pressure at liquid air temperature is so small that it is of no account.

CHAPTER IX

THE INERT GASES IN THE ATMOSPHERE

In the year 1785 Cavendish showed that when atmospheric nitrogen (phlogisticated air) was mixed with excess of oxygen (dephlogisticated air) and subjected to the prolonged action of electric sparks in presence of caustic potash, combination of the gases occurred resulting in the formation of oxides of nitrogen. The oxides of nitrogen are absorbed by the caustic potash, the result being a diminution in the bulk of the mixed gases. On absorption of the excess of oxygen by 'liver of sulphur' a small gaseous residue remained, 'which was certainly not more than $\frac{1}{120}$th part of the bulk of the phlogisticated air.'

This curious result was of the greatest interest to Cavendish, and he appears to have realised that 'phlogisticated air' was not homogeneous; at any rate his experiment seems to have raised doubts in his mind regarding the homogeneity of atmospheric nitrogen. The problem was set aside, doubtless because Cavendish's experimental methods were not sufficiently refined to investigate the phenomenon accurately.

No attention was paid to Cavendish's work, or at any rate no attempt was made to explain his result until the year 1894 when the mystery was unravelled. A series of very accurate determinations of the densities of various gases had been carried out by Lord Rayleigh who found that atmospheric nitrogen was always about one half per cent. more dense than nitrogen prepared from sources other than the atmosphere. Lord Rayleigh concluded that atmospheric nitrogen was not homogeneous, but contained a small quantity of a heavy gas. Such a conclusion was in complete accord with Cavendish's experiment. With the aid of Sir William Ramsay, Lord Rayleigh succeeded in isolating a new gas from the atmosphere.

The method which these investigators first employed consisted essentially in an improvement on Cavendish's original experiment. Atmospheric nitrogen with a considerable excess of oxygen was sparked in presence of aqueous potash confined over mercury, until the volume of gas ceased to diminish. The oxygen was then removed by means of phosphorus. This method was found to be fairly convenient for the preparation of small quantities of the inert gas, but for the preparation of large quantities a different procedure was adopted.

Metallic magnesium has long been known to be capable of combining with nitrogen at a red heat

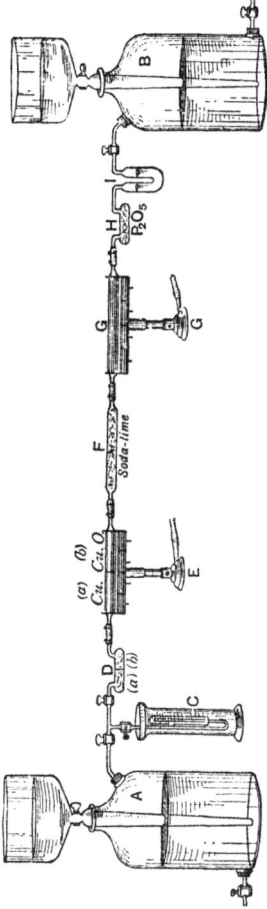

Fig. 5. Rayleigh and Ramsay's apparatus. This apparatus was employed for the preparation of argon by absorbing atmospheric nitrogen by means of heated magnesium. The nitrogen is passed from the gasholder A to the gasholder B through the tube G containing red-hot magnesium. The tube (a, b) contains copper and copper oxide for removing traces of oxygen and for oxidising any carbonaceous matter to carbon dioxide and water. The tubes F and I contain soda-lime for absorbing carbon dioxide; D and H contain phosphorus pentoxide for removing water. After passing backwards and forwards until no apparent further diminution in volume occurred, the gas is removed through the tube C and subjected to further purification by sparking with oxygen and subsequent removal of the oxygen by means of phosphorus.

forming a nitride of the metal, and Rayleigh and Ramsay found that it was a very convenient absorbent of nitrogen for the preparation of the inert gas. It was found to be difficult if not impossible to remove the last traces of nitrogen from the inert gas by this method. This was readily effected by sparking with excess of oxygen and subsequent removal of the oxygen by means of phosphorus.

Another method of absorbing atmospheric nitrogen which was found to be very effective was to employ a mixture of magnesium and lime in the proportion of three parts of the former to five of the latter. Such a mixture when heated absorbs nitrogen with great rapidity. According to Maquenne, calcium is formed by the interaction of the magnesium and calcium oxide, and this metal is the active absorbent.

The gas obtained by these methods was found to be an extraordinarily inactive substance. Treatment with the most energetic reagents failed to cause it to enter into combination. In consequence of its great chemical inactivity, its discoverers assigned to it the name of argon.

While Rayleigh and Ramsay were engaged in the investigation of argon, their attention was directed to an observation which had been made many years previously. In 1868 Janssen observed a bright spectral line (of wave length 5876 Ångström units) in the chromosphere of the sun, which was attributed

by Sir Norman Lockyer to an element existing in the
sun.   In 1889 Hillebrand observed that when certain
uranium-containing minerals were boiled with dilute
sulphuric acid, a quantity of gas was evolved, the
nature of which was not precisely determined.   In
1895 Ramsay at the suggestion of Miers investigated the gases evolved by heating the mineral
cleveite with dilute sulphuric acid *in vacuo*.   A gas
possessing several new spectral lines, including the
line which had been observed in the chromosphere
of the sun by Janssen, was isolated.   This gas was
given the name of helium in consequence of its having
been discovered in the chromosphere of the sun
before it was known on the earth.   Like argon, helium
was found to be chemically inactive.

Inasmuch as helium and argon are completely
inactive, chemical methods of determining their
atomic weights are inapplicable, and the only means
available for the determination of their atomic
weights was to determine their densities.   At this
stage the investigators were confronted with another
difficulty, viz., the question of the number of atoms
in the molecule.   Fortunately, information on this
point was readily to be obtained by a determination
of the ratio of the specific heat at constant pressure
to the specific heat at constant volume.   It had long
been known from kinetic considerations that this
ratio would approximate to the value 1·66 for a gas

having only one atom in the molecule, becoming smaller for gases with polyatomic molecules. The ratio may be readily determined experimentally by measurement of the velocity of sound in the gas. Hitherto the only monatomic gas investigated, viz., mercury vapour, was found to possess the theoretical ratio 1·66 ; but determinations of the ratio for helium and argon yielded the same number, and hence it was concluded that their molecules are monatomic also. Helium was found to have a density of 1·98 and therefore an atomic weight (which is identical with its molecular weight) of 3·96 ; while argon was found to have a density approximating to 20 and therefore an atomic weight of 40.

The discovery of these two gases naturally led Ramsay to speculate on the possibility of the existence of other gases belonging to the same group of elements. He predicted the existence of another element with an atomic weight of about 20 and two others with atomic weights higher than that of argon. With the aid of Dr Travers, he showed that the atmosphere contains no fewer than five inactive gases, and succeeded in isolating them in a pure state.

In 1898 Ramsay and Travers prepared about 15 litres of atmospheric argon in order to investigate the homogeneity of the gas. With the aid of the low temperature of liquid air, they subjected this argon to fractional distillation. A gas possessing a density

intermediate between that of helium and argon was
isolated, to which they gave the name of neon. It
was found to be a matter of the utmost difficulty to
obtain this gas free from a small quantity of helium,
and it was only obtained in a pure state by having
recourse to the extreme cold of liquid hydrogen.
Neon was found to resemble helium and argon in
possessing great chemical inertness.

The remaining two gases were isolated from the
residues obtained from the slow evaporation of liquid
air. These two gases are present in such minute
proportion in the atmosphere that they escaped
detection in the fractional distillation of 15 litres of
atmospheric argon. Ramsay and Travers found that
when liquid air is allowed to evaporate slowly, the
dregs of liquid consist almost entirely of oxygen with
small quantities of nitrogen, argon, and two heavy
gases to which they assigned the names of krypton
and xenon. The isolation of these two gases from
the atmosphere was a triumph of experimental skill,
and the thanks of all chemists are due to Sir William
Ramsay for the excellent advances which he made
in the methods of dealing with almost vanishingly
small quantities of gases. It would be impossible in
the present volume to give an adequate account of
the work of Ramsay and Travers on the isolation of
krypton and xenon ; references to a few of their
memoirs will be found in the Bibliography.

All five gases are chemically inert. They possess very characteristic spectra, and their more important physical properties indicate that they constitute a perfectly natural group of elements. Inasmuch as they are devoid of chemical properties, Ramsay places them in a new vertical column in the periodic classification, between the extreme electro-negative halogens and the extreme electro-positive alkali metals.

From the residues obtained by the evaporation of liquid air, it is possible to arrive at an approximate estimate of the proportions in which these gases exist in the atmosphere. The following numbers have been obtained by Ramsay :

> Helium, about 4 parts in 1,000,000 of air
> Neon, 1 to 2 parts in 100,000 of air
> Argon, 0·937 part in 100 of air
> Krypton, 1 part in 1,000,000 of air
> Xenon, 1 part in 20,000,000 of air

Within recent years improved methods have been devised for the separation of these gases from the atmosphere, and one of the most important advances in this direction has been made by the discovery of Sir James Dewar, to which reference was made in the preceding chapter, regarding the absorptive properties of charcoal at low temperatures. When charcoal is cooled by liquid air, it is capable of absorbing large volumes of air, and it is possible to obtain very high vacua in a remarkably short time by this method.

However the various constituents are not all absorbed by cold charcoal with the same readiness; in particular helium remains uncondensed, and neon is only absorbed to a limited extent by charcoal at liquid air temperature. A ready method of separating gaseous mixtures by fractional absorption has thus been placed in the hands of chemists, and the inert gases have been thereby isolated with much greater ease than by the original methods. Their physical properties have been more accurately studied; in particular, their density, refractivity, compressibility, critical constants, solubility in water, and thermal conductivity, are now known with a satisfactory degree of accuracy. According to the most recent determinations, the atomic weights of these gases in terms of oxygen as 16·00 are as follows :

Helium	...	...	...	...	...	3·99
Neon	...	...	...	...	...	20·2
Argon	...	...	...	...	...	39·88
Krypton	...	...	...	...	...	82·92
Xenon	...	...	...	...	...	130·2

No account of the inert gases would be complete without some reference to the connexion between helium and radioactive phenomena. For an adequate account of these phenomena, we must refer the reader to a suitable text-book. For the present purpose it is sufficient to state that radioactive substances possess the properties of conferring electrical conductivity on gases, of affecting a photographic plate and of causing certain substances to fluoresce. It is now practically certain that the source of the enormous quantities of energy liberated by these substances is the energy of disintegration of the atoms of the radioactive elements. In 1903 it was predicted by Professor Rutherford and Mr Soddy, as a result of quantitative investigations on the continuous production of radioactive matter, that helium would be one of the final products of radioactive change. In the same year Ramsay and Soddy, by spectroscopic methods, were able to detect the direct production of helium from the radium emanation, a radioactive gas which is continuously produced in extremely minute amount from the parent element. This observation has been confirmed by several other investigators. When helium was first discovered in terrestrial sources, Ramsay called attention to the fact that it is only to be found in minerals which contain uranium and thorium. These minerals are now known to be radioactive. The helium, as the result of atomic disintegration, becomes occluded in these minerals, and may be separated by heat or solution. Certain mineral springs give off relatively large quantities of helium into the atmosphere, doubtless from some radioactive source. It may be asked why the quantity of helium in our atmosphere does not tend to increase as the result of radioactive change. The answer to this question is that the quantity does tend to increase, but it will be remembered from what was stated in Chapter VII, evidence was adduced to show that this gas probably escapes from our atmosphere into space. For this reason the amount remains sensibly constant.

The extraordinary chemical inertness of these gases is naturally of great interest to the chemist. It may be asked whether, in any circumstances, these elements could possibly enter into combination. This is a question which cannot be discussed here, but it may be stated with a high degree of probability that if these elements possess any tendency towards combination, this capacity will only be exerted at high temperatures, and the compounds formed would be of a highly endothermic character. Compounds which are formed from their elements with an absorption of heat are termed endothermic, and the stability of such compounds increases with rise of temperature.

Another interesting property possessed by these gases is their relatively high resistance to the conduction of electricity at low pressures. It is well known that ordinary gases under a pressure of a few millimetres of mercury conduct the electric discharge, the particular phenomena observed depending, among other circumstances, on the pressure. In high vacua, the resistance of the residual gas becomes so great that the gas ceases to conduct the discharge. The helium gases, when carefully purified, exhibit the well known phenomena of 'fluorescent vacua' and 'non-conducting vacua' at pressures considerably higher than those at which the same phenomena are exhibited by ordinary gases. In other words, the helium gases are relatively inert electrically as well as chemically. It is probable that some connexion between the two phenomena exists. Ever since the time of Berzelius, attempts have been made with more or less success to formulate electrical theories of chemical union, and it is just possible that the inert gases may furnish an interesting subject for research in this direction.

In 1908 Sir William Ramsay instituted a search for possible new members of the inactive series of gases. As the periodic system would appear to indicate the existence of two or three elements of greater atomic weight than xenon, Moore, working in Ramsay's laboratory, fractionated the heavy gases from no less than one hundred and twenty tons of liquid air. However no positive result was obtained, and it was concluded that either the missing gases do not exist in the atmosphere, or that they dis-integrate during the process of separation. The radioactive emanations, which appear to belong to the gases of the helium group, exist in minute quantities in the atmosphere. We shall consider this subject in the next chapter.

The field of investigation which has been opened out by the discovery of the inert gases has proved to be one of very great interest, and one cannot fail to be impressed with the fact that a great deal of the success which has crowned the efforts of the workers in this field has been due to unfailing resource in the refinement of experimental methods.

CHAPTER X

THE RADIOACTIVITY OF THE ATMOSPHERE

IN Chapter IX we have referred to the connexion between helium and radioactive phenomena, and also to the search for possible new members of the group of inert gases in the residues from one hundred and twenty tons of liquid air. It was concluded from this investigation that either the missing gases do not exist in the atmosphere or that they disintegrate during the process of separation. We shall see that the latter conclusion is the more probable one.

In ordinary circumstances, dry air possesses but feeble electrical conductivity, but its conductivity is enormously increased by exposure to what have been termed 'ionising influences.' Such influences are X-rays or the rays from radioactive substances. The work of Sir J. J. Thomson and others has shown that the conductivity of gases under the influence of such sources of radiation is due to the production of positive and negative ions[1] in the gas. On removal of the ionising agency, the positive and negative ions recombine, and the gas loses its conductivity to a great extent.

The feeble electrical conductivity which the atmosphere usually exhibits was ascribed by the earlier investigators to 'leakage' in some form or another; although the nature of this leakage was in no way understood. In 1900 Mr C. T. R. Wilson and Professor Geitel independently showed, by experiments in which adequate precautions were taken to avoid defective insulation, that an electrified system, charged either positively or negatively, gradually lost its charge through the air. The leakage was the same in the dark as in diffused daylight and was independent of the locality. As precautions had been taken to ensure the absence of dust particles, it was concluded that the air was spontaneously ionised.

Now since gaseous ions once produced are continually recombining, it is evident that there must be some source of ionising energy to make good the losses. Such a source may be either the presence of traces of radioactive gases in the atmosphere or a very penetrating radiation in the soil or from space,

[1] The term *ion* was originally employed by Faraday to denote a moving particle carrying an electric charge.

which continually ionises the air in contact with it.
The ionisation of the air may be due to both sources.

In 1901 Professors Elster and Geitel attempted
to extract a radioactive substance from the air.
Rutherford had previously shown that the active
matter from the thorium emanation could be concentrated on the negative electrode in a strong electric
field, and that therefore the carriers of the radioactivity possessed a positive charge of electricity.
Elster and Geitel exposed a cylinder of wire netting
charged to a potential of 600 volts for several hours
in the open air. On removing the cylinder and placing
it in a large bell jar, inside which was placed an electroscope, it was found that the rate of discharge of the
latter was slightly increased. More successful results
were obtained by the use of a long wire exposed at
some height above the ground, and charged to a high
negative potential from one terminal of an influence
machine, the rate of discharge of the electroscope
being greatly increased. It was therefore concluded
that a radioactive substance had been extracted from
the atmosphere, and that the carriers of the radioactivity were positively charged ; since no radioactive
matter could be collected when the wire was charged
positively. The activity could be removed from the
collecting wire by rubbing it with a piece of leather
moistened with ammonia, the latter becoming strongly
radioactive, in some cases so strongly as to be capable

of affecting a photographic plate, and of causing a screen of barium platinocyanide to fluoresce.

The properties of a wire made active by exposure to the atmosphere were found to be remarkably similar to the activity imparted to a wire by exposure to the radioactive emanations of radium and thorium. Further, it was found that the rate of decay of the activity of the deposit on such a wire is approximately identical with the rate of decay of the active deposit of radium emanation. The presence of minute quantities of radium emanation in the atmosphere was thus rendered highly probable.

In 1905 Eve conducted some experiments with a view to determining the amount of radium emanation in the atmosphere. Experiments were first made with a large iron tank, 17 cubic metres in volume. The active deposit was abstracted from the air in this tank by means of an insulated wire passing through the tank and charged to a high negative potential. These experiments were carried out in a building into which no radioactive material had ever been introduced. After two hours exposure, the wire was removed and connected with a gold-leaf electroscope, and the rate of discharge measured. The rate of decay of the activity of the wire was found to be about the same as for the active deposit of the radium emanation. By comparison with experiments made with a smaller tank into which a known amount of radium emanation had been introduced previously, Eve concluded that the amount of emanation present per cubic metre of air was equivalent to the amount of emanation in equilibrium with 80×10^{-12} grams of radium.

In 1907 Eve investigated the ionisation of the atmosphere over the North Atlantic Ocean, which curiously enough appears to be of the same order as over Europe and America. The amount of radium in sea-water is quite insufficient to account for this ionisation, and is only from 1/500 to 1/1000 of the average amount in sedimentary and igneous rocks. The radium emanation in the atmosphere most probably originates from the radium in the earth, and gradually diffuses through the soil or is carried in solution by waters which percolate through the soil from which it finds its way into the atmosphere. Support to this contention is given by Sir J. J. Thomson's discovery that the Cambridge well-water contains radium emanation in solution.

In 1908 Eve determined the amount of radium emanation in the atmosphere by a different method. He passed large quantities of air over charcoal at the ordinary temperature and expelled the occluded emanation by heat. He found that the average amount per cubic metre of air was equal to that in equilibrium with 60×10^{-12} gram of radium. This quantity is of the correct order necessary to account for the amount of radium C (one of the substances giving rise to the active deposit) which can be collected on a negatively charged wire from the atmosphere. If it be assumed that 5 per cent. of the total emanation escapes into the atmosphere from the ground, it can be calculated that a depth of the earth's crust of two or three metres contains enough radium to supply the observed amount of emanation in the atmosphere to a height of ten kilometres. How then is the fact that the radioactivity of the atmosphere over the ocean is practically identical with that over the land to receive explanation? It has been stated already that the quantity of radium in sea-water is far from sufficient; and it is practically impossible for the emanation in the air over the land to diffuse uniformly over the water, since its activity rapidly decays with time, falling to half value in about 3·8 days. But, on the other hand, the quantity of radium in sea-water, although excessively minute, is *in solution*, and yields its emanation to the air. Then, too, the ever moving surface of the water, and the bubbles which rise to the surface, continually bring fresh quantities of the emanation into the air. Thus the ocean, though poor in its radium content, gives up its emanation so easily that it may be sufficient to supply all the radium emanation which the atmosphere overhead is found to contain.

Various other determinations have been made of

the amount of radium emanation in the atmosphere. The amount varies considerably with the locality and with the atmospheric conditions. It has been clearly established, however, that the amount of radium emanation is insufficient to account for more than a fraction of the observed ionisation of the atmosphere, the remainder being ascribed to a penetrating radiation from radioactive elements in the earth's crust. This penetrating radiation, which is chiefly composed of γ-rays[1], has been shown to come from the earth, since the natural rate of discharge of electroscopes is greatly reduced by enclosing them within thick lead walls which absorb the γ-radiation.

There is yet another source of the ionisation of the atmosphere. In 1907 Blanc showed, by experiments similar to those of Elster and Geitel, that 50 to 70 per cent. of the active deposit obtained from the air of Rome is due to the thorium emanation. The emanation of thorium is similar to that of radium, but its rate of decay is very much more rapid, its activity falling to half value in about 53 seconds. The air in the neighbourhood of New Haven (Connecticut) and in California also contains notable quantities of thorium emanation. It has also been

---

[1] Three main types of rays are emitted by radioactive substances which have been termed the α-, β-, and γ-rays. The last are the most penetrating; they resemble X-rays, although their nature is not yet definitely established.

recently shown that comparatively large quantities of this emanation occur in the air of Manchester. Now, since the rate of decay of the thorium emanation is extremely rapid, it follows that its origin cannot be at any great depth below the soil, but it is highly probable that the surface rocks, in places where this emanation occurs in quantity in the atmosphere, are rich in thorium.

Investigations of the radioactive emanations in the atmosphere continue to yield results of interest. Experiments carried out in 1910 at the Sestola Observatory in the Appenines at an elevation of over 3000 feet have shown that the proportion of the activity collected by a charged wire due to the thorium emanation varied considerably, being increased by a falling barometer and high winds. Satterly in 1910 determined the average amount of radium emanation in the air of Cambridge, and found an average of 1·7 molecules per cubic centimetre, but wide variations from this mean have been observed. In 1911 the same investigator carried out some measurements on the amount of radium emanation drawn up from underground, and it was found that the ground air is 2500 times as rich in radium emanation as atmospheric air.

In concluding this brief account of the natural ionisation of the atmosphere, it must be admitted that while the presence of radioactive emanations in

the atmosphere and the existence of very penetrating
emanations from the earth undoubtedly account for
many of the observed facts, the problem cannot be
said to be completely solved. In particular, the radioactivity of the atmosphere over the ocean requires
further investigation.

It has been suggested that the small quantity of
hydrogen in the atmosphere is due to the radioactive decomposition of water. This phenomenon was
discovered in 1902 by Giesel. Later experiments by
Sir William Ramsay and others showed that when
water is decomposed under the influence of the radium
emanation, an excess of hydrogen above the theoretical ratio is always obtained ; no satisfactory
explanation has been given of what becomes of the
oxygen.

In conclusion a few words must be said about the
properties of the radioactive emanations. Apart from
their radioactive properties, there is little to be
said beyond the fact that their properties indicate
that they belong to the gases of the helium group.
Most of the experiments on their chemical and
physical properties have been performed on the
radium emanation, as its comparatively long life
renders it more capable of investigation than the
emanations of thorium and actinium. These gases
are produced in such extremely minute quantity from
their parent elements, that it is safe to state that they

would never have been discovered but for their radioactive properties. Sir William Ramsay and Mr Soddy sparked the radium emanation for several hours with oxygen over caustic alkali. On removal of the oxygen, no visible residue was left. But when another gas was introduced, mixed with the minute amount of emanation in the tube, and withdrawn, its activity was found to be unaltered. All attempts to cause the emanation to enter into combination have proved fruitless. Its most remarkable property is its spontaneous change into helium. The physical properties of the emanation have been fairly completely studied. It has been found to obey Boyle's law, and to be capable of being condensed to the liquid state at low temperatures. Its solubility in various liquids has been determined, and its spectrum is now fairly well known.

For several years after the discovery of the radioactive emanations, their densities provided chemists with a very difficult experimental problem. Owing to the extremely small quantity of these emanations available for investigation, direct determination of their densities appeared to be hopeless, and consequently indirect measurements, such as their relative rates of diffusion into other gases, were resorted to. The results were far from satisfactory owing to the very large experimental errors involved ; but indications of a molecular weight of the order of 200 were

obtained for the radium emanation. However, in
1909, Steele and Grant constructed a quartz micro-
balance capable of detecting changes of weight if
greater than 4×10^{-9} gram, and in the following year
Sir William Ramsay and Mr R. W. Gray employed
this balance to determine the density of the emanation.
Although the quantities of emanation were almost
infinitesimal, in no case exceeding 0·1 cubic millimetre
at atmospheric pressure, fairly consistent results
were obtained for the molecular weight of the
emanation. The extreme values found were 216 and
228, with a mean of 220. From radioactive considerations, the atomic weight of the radium emanation
should be 222·6, being four units less than that of
radium (226·6) due to the loss of a single alpha particle,
which has been proved to be identical with a helium
atom. The fact that a value so close has been obtained by direct weighing would have been incredible
a few years ago.

The radium emanation falls naturally into the
vertical group of inert gases in the periodic classification two places below xenon. Ramsay and Gray
propose to assign to the gas the name of niton. So
far as their chemical behaviour is concerned, there is
no doubt that the emanations of the other two radioactive elements, thorium and actinium, belong to the
same group of elements, and we may hope that with
the advent of more refined methods, their atomic

weights will be accurately known and the column in
the periodic table be completed.

It may be pointed out that in one particular
direction, further investigation of the radioactivity of
the atmosphere may be of the greatest importance.
Within recent years, numerous investigations of the
properties of the spas and thermal springs in various
parts of the world have been made on account of the
supposed medicinal value of the dissolved radium
emanation. If it should be found that the health-
giving properties of the atmosphere of any particular
locality are dependent on the presence of traces of
radium emanation, the importance of further research
in this direction can scarcely be over-estimated.

Closely connected with the subject of the ionisation of the atmosphere is the much older one of
atmospheric electricity, which has formed a subject
for investigation ever since the identity of the light-
ning flash and the electric spark was demonstrated
in the eighteenth century by Benjamin Franklin.
Without going into details in regard to the earlier
investigators who performed many interesting experiments in collecting electricity from the air, mention
must be made of the work of Volta, who employed a
slow burning match attached to a long metal rod. The
electricity in the air in the vicinity of the flame induces
electricity of the opposite sign at the upper end of the
rod which is constantly being carried off by convection

currents in the flame, with the result that the con-
ductor becomes charged with electricity of the same
sign as the atmosphere. A modification of Volta's experiment was made at a later date by Lord Kelvin, who
employed a water-dropping apparatus for determining
the electrical potential of the air at any point. Kelvin
employed an insulated reservoir filled with water
which issued drop by drop from a small orifice. In
a very short time, the can was found to be electrified
to the same extent as the air around it, and the
potential could be determined by means of a quadrant
electrometer. In frosty weather a slow burning
match was employed in conjunction with the electrometer.

The results of these and other experiments have
shown that under a clear sky the atmosphere is nearly
always electrified positively. But in certain circumstances the sign of the electrification of some parts of
the air may become negative, and then thunderstorms
may be expected.

It has been already explained that the ionisation
of the atmosphere is due, in part at any rate, to the
presence of radioactive emanations. But air may be
ionised in other ways, such as by ultra-violet light of
extremely short wave-length, by high temperatures,
and by certain chemical processes ; it is, however,
extremely doubtful if any of these processes contribute
to the natural ionisation of the atmosphere.

The researches of C. T. R. Wilson and of Elster
and Geitel have shown that when charged conductors
lose their charge in dust-free air, the rate of dissipation of charge is greater when the conductors are
charged negatively than when they are charged
positively. The reason of this is that the lower
atmosphere contains more positive than negative ions.
It has also been shown that there is a connexion
between potential gradient and the rate of dissipation
of electric charge; when the former is high the latter
is low and *vice versâ*. No satisfactory explanation
of the earth's negative charge has been given, although
Wilson thinks that it may be partly accounted for by
the condensation of aqueous vapour on ions which takes
place more readily on negative ions than on positive
ions. Rain-drops thus formed, charged with negative
electricity, will give up their charge to the earth on
arriving at its surface, but Wilson expressly states
that although the earth may receive its negative
charge in this way, some other explanation must be
sought for the maintenance of the charge.

In an interesting *résumé* of the normal electrical
phenomena of the atmosphere published in 1905 by
Mr G. C. Simpson, it was stated that the discovery
of ions and of atmospheric radioactivity instead of
solving the chief problem of atmospheric electricity,
has only made the difficulties to be explained still
more formidable. It appears also that it is extremely

doubtful if there is any connexion between the normal conditions of atmospheric electricity and the aurora[1]. After spending a year in the region of maximum aurora, and experimenting with the best instruments available, he failed to detect the slightest connexion between the phenomena of the aurora and the electrical conditions of the lower atmosphere.

CHAPTER XI

THE PROBABLE COMPOSITION OF THE ATMOSPHERE IN EARLY GEOLOGICAL TIME

In this concluding chapter we desire to call attention to some of the views which have been held regarding the composition of the atmosphere in primitive times. It has been held by some geologists

[1] It is interesting to note that the spectrum of the aurora possesses a green line of identical wave-length with the green line of krypton ($\lambda = 5570$). This has given rise to the suggestion that the aurora is an electric discharge phenomenon in which krypton is concerned. The spectrum of the aurora has recently (1912) been investigated with the object of verifying or disproving this suggestion, and it has been found that the yellow krypton line ($\lambda = 5871$) does not exist in the auroral spectrum. Further, it has been shown that alterations of the conditions of the electric discharge in krypton do not produce any alteration of the relative intensities of the two lines. It would appear, therefore, that krypton is not concerned in the production of the phenomena of the aurora.

that before and after the Carboniferous period the atmosphere was more highly charged with carbon dioxide than it is at the present time.   It has been suggested that during the growth of the Coal period forests the atmosphere became enriched with oxygen. On the other hand, the production of large quantities of carbon dioxide by combustion and volcanic activity has led some to believe that the proportion of this constituent is increasing, while that of the oxygen is decreasing.   About sixty years ago, Koene enunciated the hypothesis that the earth's primitive atmosphere consisted of nitrogen, carbon dioxide, and aqueous vapour, but contained no free oxygen, basing his views on the fact that the older rocks contain oxidisable matter.   Similar views were expressed at a much later date by Phipson.   In 1897 Lord Kelvin delivered his famous address on the Age of the Earth, and adduced evidence to show that the primitive atmosphere was devoid of oxygen, although his opinion on this matter does not appear to have been very definite.

Between 1900 and 1906 Mr J. Stevenson, in an interesting series of papers on the chemical and geological history of the atmosphere, arrived at the same conclusion, viz., that there was a time when there was no free oxygen on the earth.   The data on which he reasoned were principally the probable amount of coal and other carbonaceous matter in

the earth, and the rate of growth of vegetation.
Following this line of reasoning, he pointed out that
there is a deficiency of oxygen on the earth relative
to the total amount of oxidisable matter. But, as
he remarked, it does not follow that there was ever
a time when the earth's atmosphere was totally devoid
of oxygen. At a time when the temperature of the
earth was very high, probably above the temperatures
at which most oxides are dissociated, there might
have been a considerable quantity of free oxygen.
As the temperature of the earth fell, the oxygen
would unite with various elements, and probably
large quantities of silicates would be formed. These
silicates would probably form a protecting envelope,
and prevent the union of free oxygen with elements
lower down.

But the difficulties in the way of accepting any
such hypothesis are very great. The ratio of our
total free oxygen to the total combined oxygen is
exceedingly small, and as considerable quantities of
oxidisable matter still exist in the earth, it is difficult
to understand why there should be any free oxygen
remaining at all, were it not that some counteracting
influence is at work. According to Stevenson, it is
much more probable that the oxygen was all in the
combined condition at one time ; and further, that
the atmosphere contained hydrogen and hydrocarbons
after all the oxygen had entered into combination ;

and that our present supply of oxygen has been produced by the decomposition of carbon dioxide by plants under the influence of sunlight.

It has been shown that certain forms of vegetation can flourish in atmospheres of hydrogen or certain hydrocarbons. After the death of such vegetation, the carbon on becoming heated in contact with iron oxide would form metallic iron and carbon dioxide. Such may have been the origin of the metallic iron found in the earth and of the carbon dioxide in the atmosphere. Once the carbon dioxide had been produced, the production of oxygen by vegetable activity would follow as a necessary consequence. The probability of the truth of any such hypothesis of the origin of atmospheric carbon dioxide is, however, dependent to some extent on our knowledge of the first forms of vegetable life on this once molten planet. Such vegetation must have consisted of organisms of the simplest type.

The very high temperature of the primitive earth must have effected the dissociation of limestones and other naturally occurring carbonates, with the result that the then existing atmosphere must in all probability have contained large quantities of carbon dioxide. The vast quantities of such naturally occurring carbonates have given rise to estimates that the primitive atmosphere must have been some two hundred times greater in extent than the present one.

In a previous chapter we have indicated the causes which tend to increase the quantity of carbon dioxide in the atmosphere and also those which tend to decrease its amount, and have pointed out that the amount remains fairly constant within narrow limits. It is interesting to speculate on the question as to whether, granting that the earth's primitive atmosphere was rich in this constituent, the amount of carbon dioxide has diminished steadily until the present minute proportion (3 volumes in 10,000 volumes of air) has been reached, or whether, during various geological epochs, the amount has alternately increased and diminished, according to the conditions which have prevailed. The latter view was taken by Stevenson, and some of the evidence which he adduced in favour of it may be summarised here.

Inasmuch as volcanic action is one of the most important sources of this gas in the atmosphere, it is reasonable to assume that the amount of atmospheric carbon dioxide was greater in those geological epochs which were characterised by great volcanic activity. On the other hand, it is highly probable that in the event of a large increase in the percentage of atmospheric carbon dioxide, there would be a concomitant increase in the growth of a luxuriant vegetation over the earth, other things being equal. But, as was pointed out by Stevenson, it is probable that if the luxuriance of vegetation were to be greatly

increased the quantity of carbon dioxide would tend
to be increased also, as a result of the oxidation of the
vegetable remains, and that to an extent probably
not much less than the ratio of the increase in the
luxuriance itself. He concluded that the amount
of carbon dioxide has varied over wide limits, limits
sufficiently wide, when regard is taken for the high
absorptive power of this gas for radiant heat to
account for the great climatic changes of geological
history.

We must now refer to Stevenson's attempt to
answer the question. What is the state of matters
in the above respect at the present time? Is the
amount of carbon dioxide increasing or decreasing,
or has it been practically constant for many years,
and has it still a tendency to remain at the same
figure? He pointed out quite correctly that little is
to be gained by a study of the records of the older
analyses in comparison with those which have been
made in more recent years, as the older determinations are not sufficiently accurate.

We return to the consideration of the functions
of the waters of the ocean as a regulator of the
amount of carbon dioxide in the atmosphere to
which reference was made in Chapter IV. According
to Stevenson, the amount of available carbon dioxide
in the sea as estimated by Schloesing is probably too
high, basing his contention on the fact that the

larger proportion of this gas in the sea is in the
combined forms of calcium bicarbonate and carbonate,
and therefore not in a form readily available for
interchange with the atmosphere. In support of
this contention he quoted the results of Dittmar's
analyses made in the course of the *Challenger*
Expedition. Dittmar found that the sea rarely contained any free carbon dioxide, as there was usually
enough to form normal calcium carbonate although
an insufficient quantity to form the bicarbonate,
and the reaction of sea-water was usually alkaline.
Stevenson considered that we should require to have
much more knowledge than we possess at present
as regards the various factors which determine the
amount of carbon dioxide in sea-water. In particular,
we should require to have some knowledge regarding
the amount of this gas produced by the slow oxidation
of organic matter in the sea, and that produced by
the respiration of marine animals. We should also
require to know more about the removal of carbon
dioxide from the sea by the assimilatory activity
of marine algae, and its removal by the action of
sea-water on rocks containing basic material. The
analyses of various observers appear to indicate that
the amount of carbon dioxide in the atmosphere is
slightly greater over the land than over the ocean,
although we should need many more analyses than
we possess at present before making a dogmatic

statement. However, if this constituent is more
abundant over the land, then on the hypothesis
that the ocean is an efficient regulator, Stevenson
reasoned that the amount of carbon dioxide is
increasing at the present time.

Various factors may contribute to this increase,
and there can be little doubt that the increased
consumption of coal has some influence. Stevenson
estimated that in the next twenty years the increased
consumption of coal would have an effect in increasing
the carbon dioxide in the atmosphere to an extent
sufficient to be detected by chemical analysis, and
further, that in the next hundred years the increase
in the amount of this constituent would be sufficient
to affect the climate or average temperature of the
surface of the earth.

In conclusion, we may refer to Stevenson's views
on the influence of the variations of the amount of
carbon dioxide in the atmosphere on the rate of the
secular cooling of the earth. As is well known, the
opinions of physicists and of biologists differ considerably on this matter. Lord Kelvin's estimate,
based on physical data, assigned from ten to twenty
million years as the maximum age of the earth as
a habitable world. But many biologists and geologists
postulate a period of several hundred million years
for the evolution of the inorganic and organic world
as it exists to-day. As Stevenson has pointed out,

if it be true that the earth in former geological
epochs possessed an atmosphere much richer in
carbon dioxide than the present one, and more
especially if that atmosphere was much greater in
extent, our knowledge of the high absorptive power
of carbon dioxide for radiant heat surely points to
the fact that Kelvin's estimate of the rate of the
secular cooling of the earth is too high. In other
words, the probable previous history of the atmosphere may be adduced as evidence in favour of the
geological estimate of the age of the earth.

Radioactive phenomena appear to lead to the
same conclusion. We shall, however, mention only
one of the lines of enquiry in this direction. During
the last four or five years the Hon. R. J. Strutt made
some determinations of the accumulation of helium
in geological time. A number of minerals have been
examined and an approximate estimate has been
made of their geological age from the helium content.
Such an estimate is necessarily a minimum estimate, as
there can be little doubt that some of the helium
must have escaped. In many cases, however, a very
fair agreement has been found between the amount
of helium and the amount of uranium or thorium in
the mineral. There are, however, one or two excep-
tions to this; the minerals beryl and tourmaline
contain notable quantities of helium and practically
no radioactive matter. Minerals such as the zircons,

9—5

which contain hundreds of times as much helium as the rocks surrounding them, must have been generated since the consolidation of the rock and the separation of the mineral. Strutt assigned minimum ages of 250 and 280 million years to two specimens of thorianite, as the result of the direct determination of the helium content, and its calculated rate of production from the radioactive constituents of the mineral.

A question which is of great interest at the present time is this : is the small quantity of helium in our atmosphere (four parts per million) entirely of radioactive origin ? It has been stated already that helium probably escapes from our atmosphere into space, but that the large sources of this gas from mineral springs in various parts of the world tend to keep the amount of this constituent of the atmosphere sensibly constant. It is difficult to imagine means of arriving at any idea of the rate at which helium escapes from minerals in which it is occluded. It was shown a few years ago by Jaquerod that helium is capable of diffusing through quartz at temperatures above about 500° C. There can be little doubt that if from any cause a mineral became heated after its formation, for example by volcanic activity, some of the occluded helium would probably diffuse through the siliceous matter and thus find its way into the atmosphere. The atmosphere may thus have become enriched with helium in times of great volcanic activity. On the other hand, very little is known regarding the *quantity* of helium which escapes from the atmosphere owing to the acceleration of the translatory motion of the molecules beyond the critical velocity. As Stoney has pointed out, the application of Maxwell's law of the distribution of velocities is incorrect; some totally different law would have to be applied to the conditions which prevail at the outer limits of our atmosphere. The previous history of the helium in our atmosphere is thus to be added to the many unsolved problems which await investigation.

BIBLIOGRAPHY

No attempt at completeness has been made in compiling the following list of works. References are restricted to books and original papers which are of particular interest or in which certain topics are treated in detail.

(1) Von Meyer, Ernst. A History of Chemistry from the earliest times to the present day. Translated with the author's sanction by George McGowan. London, 1906.

(2) Thorpe, Sir Edward. Essays in Historical Chemistry. Third Edition. London, 1911.

(3) Roscoe, Sir H. E. and Schorlemmer, C. A complete Treatise on Inorganic and Organic Chemistry. Vol. I. The Non-Metallic Elements. Fourth Edition revised by Sir Henry Roscoe assisted by Dr J. C. Cain. London, 1911.

(4) Strutt, Hon. R. J. A Chemically Active Modification of Nitrogen produced by the Electric Discharge. (Roy. Soc. Proc. 1911, 85 A, p. 219 ; Ibid. 1911, 86 A, p. 56 ; Ibid. 1912, 86 A, p. 262 ; Ibid. 1912, 87 A, p. 179.)

(5) Bone, William Arthur. On Gaseous Combustion. British Association Report, 1910.

(6) Bodländer, G. Ueber langsame Verbrennung. Stuttgart, 1899.

(7) Dunstan, W. R., Jowett, H. A. D. and Goulding, E. The Rusting of Iron. (Trans. Chem. Soc. 1905, p. 1548.)

(8) Moody, G. T. (Ibid. 1906, p. 720.)

(9) Tilden, Sir William. (Ibid. 1908, p. 1356.)

(10) Lambert, B. and Thomson, J. C. The Wet Oxidation of Metals. Part I. The Rusting of Iron. (Ibid. 1910, p. 2426.) Part II. The Rusting of Iron (continued). (Ibid. 1912, p. 2056.)

(11) Dunstan, W. R. and Hill, J. R. (Ibid. 1911, p. 1835.)

(12) Smith, R. Angus. On Air and Rain. London, 1872.

(13) Wegener, A. Die Erforschung der obersten Atmosphären-
schichten. (Zeitsch. anorg. Chem. 1912, 75, p. 107.)

(14) Stoney, G. Johnstone. The Escape of Gases from Planetary
Atmospheres according to the Kinetic Theory. (Astrophysical Journal, VII. 1898, p. 25 : XI. 1900, pp. 25 and
357 : XII. 1900, p. 201.)

(15) Cook, S. R. (Ibid. XI. 1900, p. 36.)

(16) Bryan, G. H. The Kinetic Theory of Planetary Atmospheres.
(Roy. Soc. Proc. 1900, 66, p. 335.)

(17) Rogovsky, E. On the Temperature and Composition of the
Atmospheres of the Planets and the Sun. (Astrophysical
Journal, XIV. 1901, p. 234.)

(18) Travers, M. W. The Experimental Study of Gases. London,
1901.

(19) Rayleigh, Lord and Ramsay, W. Argon a new Constituent
of the Atmosphere. (Phil. Trans. 1895, 186 A, p. 187.)

(20) Ramsay, W. and Travers, M. W. The Preparation and some
of the Properties of Pure Argon. (Roy. Soc. Proc. 1898,
64, p. 183.)

(21) Ramsay, W. and Travers, M. W. Argon and its Companions.
(Phil. Trans. 1901, 197 A, p. 47.)

(22) Ramsay, Sir William. Search for Possible New Members of
the Inactive Series of Gases. (Roy. Soc. Proc. 1908, 81 A,
p. 178.)

(23) Moore, Richard B. Heavy Constituents of the Atmosphere.
(Roy. Soc. Proc. 1908, 81 A, p. 195.)

(24) Rutherford, E. Radioactivity. Second Edition. Cambridge,
1905.

(25) Joly, J. Radioactivity and Geology. London, 1909.

(26) Soddy, F. Annual Reports of the Chemical Society. 1904
to 1911.

(27) Simpson, G. C.   Normal Electrical Phenomena of the Atmosphere.   (Quart. Journ. Roy. Meteor. Soc. 1905, XXXI. p. 295.)

(28) Stevenson, J.   The Chemical and Geological History of the Atmosphere.   (Phil. Mag. 1900, v. Vol. 50, pp. 312 and 399 ; Ibid. 1905, VI. Vol. 9, p. 88 ; Ibid. 1906, VI. Vol. 11, p. 226.)

(29) Strutt, Hon. R. J.   The Accumulation of Helium in Geological Time.   (Roy. Soc. Proc. 1908, 81 A, p. 272 ; Ibid. 1909, 83 A, p. 96 ; Ibid. 1910, 83 A, p. 298 ; Ibid. 1910, 84 A, p. 194.)

(30) Clarke, F. W.   The Data of Geochemistry.   (Bulletin, U. S. Geological Survey.   2nd Edition, 1911.)

INDEX

Acids, Lavoisier's theory of the composition of, 25
Age of the Earth, 130, 136, 137
Air pump, 4, 5
Aldehydes, 64, 65
Allotropy, 35, 42
Ammonia in the atmosphere, 53
Anaxagoras, 1
Anderlini, 86
Andrews, 43, 94
Argon, 99, 106, 107, 108, 109, 110, 111
Armstrong, H. E., 62, 63, 74, 75
Atom, development of the conception of, 28, 29, 30, 31
Aurora, 129
Autoxidation, 66

Baeyer, 48
Baker, H. B., 61, 63
Baly, 100
Barometer, 9, 10
Becher, 12
Beijerinck, 37
Bernoulli, Daniel, 87
Berthelot, 58
Berthollet, 29
Berzelius, 30, 42, 114
Birkeland, 39
Black, Joseph, 13, 16, 17, 22, 47
Blagden, 27
Blanc, 121
Bodländer, 67

Boltzmann, 87
Bone, W. A., 64, 65
Boussingault, 79, 80
Boyle, Robert, 5, 6, 7, 8, 11, 13, 16, 21
Bradley, 39
Brand, 21
Brown, Crum, 67
Bryan, G. H., 92
Bunsen, 58, 78

Cagniard de la Tour, 94
Cailletet, 95, 96
Calvert, Crace, 67
Carbon assimilation, 47
Carbon dioxide, 46 et seq.
Carboniferous period, 49, 130
Cavendish, 18, 19, 22, 26, 27, 28, 76, 77, 78, 103, 104
Charcoal, absorption of gases by, 101, 110, 111
Charles' law, 8
Clausius, 87
Combustion, Lavoisier's theory of, 25
Combustion, modern views on, 56 et seq.
Cook, S. R., 91
Cork, Earl of, 5
Coronium, 85, 86
Coward, 59
Critical temperature, 94
Curtius, 48

Dalton, John, 8, 29, 30, 31
Davy, Sir H., 57, 58, 78
Density of the inert gases, 107
Density of radium emanation, 124, 125
Dewar, Sir J., 41, 96, 97, 98, 101, 102, 110
Dittmar, 135
Dixon, H. B., 58, 59, 60, 61, 62
Döbereiner, 32
Dulong, 32
Dunstan, W. R., 68, 71, 72, 73

Electricity, atmospheric, 126, 127, 128, 129
Elster, 117, 121, 128
Empedocles, 1
Engler, 67
Escape of gases from planetary atmospheres, 86 et seq.
Eve, 118, 119
Eyde, 39

Falk, 59, 60
Faraday, 93, 94
Fenton, 48
Fixation of nitrogen, 38, 39
Fleming, 101
Formaldehyde, 48, 54
Frankland, Sir E., 79, 82
Franklin, Benjamin, 126
Franzen, 48

Galileo, 1, 2
Gautier, 54
Gay-Lussac, 8, 78
Geitel, 116, 117, 121, 128
Geocoronium, 85, 86
Giesel, 123
Gmelin, 32
Goulding, E., 68
Gray, R. W., 125

Guericke, Otto von, 3, 4, 5, 11

Hampson, 97
Helium, 84, 85, 86, 90, 93, 98, 100, 107, 108, 109, 110, 111, 112, 114, 115, 124, 137, 138, 139
Hellriegel, 37
Helmont, Van, 16
Henriet, 54
Hill, J. R., 73
Hillebrand, 107
Humidity of the atmosphere, 52, 53
Hydrogen in the atmosphere, 54, 84, 85, 123
Hydrogen, liquefaction of, 97, 98
Hygrometers, 83

Ignition point, 58, 59, 60, 61
Inert gases in the atmosphere, 56, 103 et seq.
Ionisation of the atmosphere, 116 et seq.
Iron, rusting of, 67–75

Janssen, 106, 107
Jaquerod, 138
Joule, 96
Joule-Thomson effect, 96, 97
Jowett, H. A. D., 68
Jupiter, atmosphere of, 91

Kelvin, Lord, 96, 127, 130, 136, 137
Kinetic theory, 86 et seq.
Koene, 130
Krönig, 87
Krypton, 109, 110, 111, 129
Kunkel, 21

Lambert, 73, 74, 75

Lavoisier, 22, 23, 24, 25, 26, 27, 28, 33, 47, 57
Le Chatelier, 58, 62
Leduc, 42, 78
Leguminosae, assimilation of nitrogen by the, 37
Linde, 97
Liquid air, 93 et seq.
Lockyer, 107
Lovejoy, 39
Low temperatures, properties of matter at, 101, 102

Magdeburg hemispheres, 5
Magnesium, absorption of nitrogen by, 104, 105, 106
Mallard, 58
Manchot, 67
Maquenne, 106
Mariotte, 7
Mars, atmosphere of, 91
Maxwell, Clerk, 18, 87, 91, 92, 139
Mayow, John, 11, 24
Mendeleeff, 32, 85, 86
Mercury, atmosphere of, 91
Meyer, Lothar, 32
Miers, 107
Minerals, geological age of, 137, 138
Moissan, 102
Molecule, development of the conception of, 28, 29, 30, 31
Moody, 68, 69, 70, 71, 72, 73, 74
Moore, R. B., 114
Morley, 42, 78

Nasini, 86
Neptune, atmosphere of, 91
Nernst, 59
Newlands, 32

Nitrification, 37
Nitrogen, 33 et seq.

Olszewski, 96
Onnes, Kamerlingh, 96, 98, 101
Optically pure air, 55
Organic matter in the atmosphere, 55
Oxygen, 40 et seq.
Ozone, 42, 43, 44, 45, 46

Paracelsus, 16
Pascal, 3
Pasteur, 56
Peligot, 49
Perier, 3
Periodic law, 32
Permanent gases, 94
Petit, 32
Pettenkofer, 81
Phipson, 130
Phlogistic period, 11 et seq.
Photosynthesis, 47, 48
Pictet, 95, 96
Planetary atmospheres, escape of gases from, 86 et seq.
Plants, assimilation of carbon by, 47, 48
Plants, assimilation of nitrogen by, 36, 37
Priestley, Joseph, 13, 14, 15, 22, 24, 26, 27, 28
Primitive atmosphere, 129 et seq.
Proust, 29

Quartz, diffusion of helium through, 138

Radioactivity of the atmosphere, 115 et seq.
Radium emanation in the atmosphere, 118 et seq.

Ramsay, Sir W., 104, 105, 106, 107, 108, 109, 110, 112
Rayleigh, Lord, 42, 54, 104, 105, 106
Regnault, 51, 78
Rusting of iron, 67 et seq.
Rutherford, 17, 18
Rutherford, E., 112, 117

Safety lamp, 57
Salvadori, 86
Satterly, 122
Saturn, atmosphere of, 91
Scheele, 13, 14, 17, 21
Schloesing, 50, 134
Schönbein, 42, 66
Scott, A., 42
Sea-water, regulator of atmospheric carbon dioxide, 50, 134, 135
Sea-water, radium emanation in, 119, 120
Shelburne, Lord, 24
Simpson, G. C., 128
Smith, Angus, 78, 79, 82
Soddy, F., 112, 124
Soret, 45, 46
Stahl, 12, 13
Standard atmospheric pressure, 9
Stevenson, J., 130, 131, 133, 134, 135, 136
Stoney, G. Johnstone, 86, 88, 89, 90, 91, 92, 93, 139
Strutt, Hon. R. J., 35, 36, 137, 138
Symbiosis, 37

Terra pinguis, 12

Thilorier, 94
Thomson, J. C., 73, 74
Thomson, Sir J. J., 115, 119
Thorium emanation, 121, 122, 123, 125
Tilden, Sir W., 72
Torricelli, 2, 3, 4, 9
Traube, M., 67
Travers, 98, 108, 109
Tyndall, 55

Uranus, atmosphere of, 91

Vacuum, nature's abhorrence of, 1, 2
Vacuum vessels, 98, 99
Van Marum, 42
Velocity of a gas molecule, 88
Venus, atmosphere of, 91
Viviani, 2
Volta, 26, 126, 127

Warltire, John, 26
Water, catalytic action of, 61, 62, 63
Water, compound nature of, 26, 27
Water vapour in the atmosphere, 51, 52, 53
Waterston, 87
Watt, James, 27
Wegener, 84, 85
Weight of the air, Galileo's attempt to measure, 2
Wilfarth, 37
Wilson, C. T. R., 116, 128
Wroblewski, 96

Xenon, 109, 110, 111

For EU product safety concerns, contact us at Calle de José Abascal, 56–1°,
28003 Madrid, Spain or eugpsr@cambridge.org.

www.ingramcontent.com/pod-product-compliance
Ingram Content Group UK Ltd.
Pitfield, Milton Keynes, MK11 3LW, UK
UKHW010851090126
466816UK00011B/151